Cost Accounting and Financial Management for Construction Project Managers

Len Holm

Routledge
Taylor & Francis Group

LONDON AND NEW YORK

The following documents are available on the eResource for this book: www.routledge.com/ Cost-Accounting-and-Financial-Management-for-Construction-Project-Managers/Holm/p/book/ 9781138550650

This first list is made available to students and instructors:

- Expanded and alternate versions of some figures utilized within the book:
  - Figure 4.3Alt: Concrete QTO for concrete COP
  - Figure 4.4Alt: Concrete pricing recapitulation sheet for COP
  - Figure 4.4L: Detailed construction estimate
  - Figure 4.5Alt: Summary estimate
  - Figure 5.1L: Detailed construction schedule
  - Figure 5.2L: Detailed jobsite general conditions estimate
  - Figure 6.1L: Detailed home office general conditions estimate
  - Table 10.1Alt: Overhead spread by specialty
  - Table 13.1L: Monthly cost loaded schedule
  - Table 14.3BL: Blank live SOV continuation spreadsheet template
  - Figure 15.1Alt: Concrete change order proposal
  - Figure 19.2Live: Live Excel pro forma
- Other documents referred to within the book:
  - Case study organization chart, reference Chapter 1 and others
  - AIA A102 contract instructions
  - Sample AIA A102 contract, reference Chapter 3 and others
  - Jobsite layout plan, reference Chapter 11
  - Expanded table of contents
  - Live Excel estimating forms utilized in Chapter 4 of this text and in *Construction Cost Estimating, Process and Practices*, Pearson, 2005.

The following are also made available to instructors:

- Instructor's Manual, complete with answers to all of the review questions and many of the exercises. A select group of case studies from *Who Done It? 101 Case Studies in Construction Management* (Amazon/Create Space, 2017), which will be published in a new format under the title *101 Case Studies in Construction Management* by Routledge in 2019, are also included.
- In addition to many exercise solutions included in the instructor's manual, several separate spreadsheet files are also provided, including: Exercises 1.2/3.6, 1.7/3.2, 2.9/10, 10.5, 12.7, 12.9, 13.8/9, 14.2, 15.4, 16.3, 18.2/3, 19.4, and others
- PowerPoint lecture slides for all 19 chapters, 19 separate files, almost 680 slides in total.

Cost Accounting and Financial Management for Construction Project Managers

Proper cost accounting and financial management are essential elements of any successful construction job, and therefore make up essential skills for construction project managers and project engineers. Many textbooks on the market focus on the theoretical principles of accounting and finance required for head office staff like the chief financial officer (CFO) of a construction firm. This book's unique practical approach focuses on the activities of the construction management team, including the project manager, superintendent, project engineer, and jobsite cost engineers and cost accountants. In short, this book provides a seamless connection between cost accounting and construction project management from the construction management practitioner's perspective.

Following a complete accounting cycle, from the original estimate through cost controls to financial close-out, the book makes use of one commercial construction project case study throughout. It covers key topics like financial statements, ratios, cost control, earned value, equipment depreciation, cash flow, and pay requests. But unlike other texts, this book also covers additional financial responsibilities such as cost estimates, change orders, and project close-out.

Also included are more advanced accounting and financial topics such as supply chain management, activity-based accounting, lean construction techniques, taxes, and the developer's pro forma. Each chapter contains review questions and applied exercises and the book is supplemented with an eResource with instructor manual, estimates and schedules, further cases and figures from the book.

This textbook is ideal for use in all cost accounting and financial management classes on both undergraduate and graduate level construction management or construction engineering programs.

Len Holm is a senior lecturer in construction management at the University of Washington, USA, and a construction professional with over 40 years' experience in various roles, including supervision of project managers, estimators, and jobsite cost accountants. He runs his own construction management firm, has developed and taught several new courses for construction management students at the University of Washington, and is the author of numerous textbooks, including *Management of Construction Projects*, second edition, with John Schaufelberger, and *Introduction to Construction Project Engineering*, with Giovanni Migliaccio, also published by Routledge.

First published 2019
by Routledge
2 Park Square, Milton Park, Abingdon, Oxon OX14 4RN

and by Routledge
711 Third Avenue, New York, NY 10017

Routledge is an imprint of the Taylor & Francis Group, an informa business

© 2019 Len Holm

British Library Cataloguing-in-Publication Data
A catalogue record for this book is available from the British Library

Library of Congress Cataloging-in-Publication Data
Names: Holm, Len, author.
Title: Cost accounting and financial management for construction project managers / Len Holm.
Description: Abington, Oxon ; New York, NY : Routledge, 2018. | Includes index.
Identifiers: LCCN 2018011969| ISBN 9781138550643 (hardback : alk. paper) | ISBN 9781138550650 (pbk. : alk. paper) | ISBN 9781315147307 (ebook)
Subjects: LCSH: Building—Estimates. | Construction projects. | Cost accounting. | Construction industry—Costs. | Project management.
Classification: LCC TH435 .H85 2018 | DDC 690.068/1—dc23
LC record available at https://lccn.loc.gov/2018011969

ISBN: 978-1-138-55064-3 (hbk)
ISBN: 978-1-138-55065-0 (pbk)
ISBN: 978-1-315-14730-7 (ebk)

Typeset in Sabon
by Keystroke, Neville Lodge, Tettenhall, Wolverhampton

Visit the eResource: www.routledge.com/Cost-Accounting-and-Financial-Management-for-Construction-Project-Managers/Holm/p/book/9781138550650

Contents

Figures

Tables

Preface

All of the construction cost accounting textbooks available are focused on the role of the chief financial officer (CFO) and chief executive officer (CEO) and home office financial management biased aspects. Very few construction management (CM) university graduates will become CFOs or certified public accountants, and although some aspire to become CEOs, those opportunities will only be available for a select few, and will occur many years after graduation. Many CM students already have some practical construction internship experience and their exposure has primarily been out on jobsites. After graduation most will begin their careers as jobsite project engineers or in home office staff support roles such as assistant estimators. Many will achieve their seven- to ten-year goal of becoming construction project managers (PM) and being placed in charge of all operations at the jobsite, especially project financial management. These students have a difficult time connecting with college textbooks focused solely on home office accounting. It is because of their work experiences and the difficulty they have in connecting with traditional cost accounting texts that I routinely have supplemented a standard accounting text with many jobsite financial management topics borrowed from my other estimating, cost control, and project management resources.

Construction project managers are not accountants, but most of what we do is accounting-related. The focus of this book therefore is on the 'Cost Accounting and Financial Management of Construction Project Managers' *at the jobsite level* and the relationship between jobsite financial management and the home office accounting department. The PM is responsible to report to the CFO and CEO for all financial affairs that happen at the jobsite, including estimating, cost control, equipment charges, cash flow, pay requests, change orders, close-out, and many others. These therefore are the financial management subjects this book couples with traditional construction cost accounting topics.

The jobsite financial management team includes the project manager, superintendent, project engineer, and if applicable, a cost engineer and/or jobsite accountant. The construction team manages jobsite general conditions and the home office executive team, including the CFO and

CEO, manages home office general conditions, and establishes profit goals for the company. This book also connects with the cost accounting activities performed by the home office including differences between alternative corporate structures, development of financial statements, equipment depreciation, and taxes and audits. Advanced financial management aspects of earned value, activity-based costing, lean construction techniques, value engineering, supply chain material management, and time value of money are also introduced for the ambitious construction cost accounting student. The pinnacle of this study concludes with a discussion of the real estate developer, who is often the general contractor's client; including the creation and management of the developer's cost pro forma model, of which construction cost is only one element.

To supplement typical academic coverage of construction cost accounting, this book includes a practical construction perspective stemming not only from my 40 years of construction experience, but from input of many construction professionals and friends. These practitioners have reviewed chapter drafts and provided input to countless figures, tables, and exercises. Their experience is very much appreciated, for without them this would just be another college textbook. It would be difficult to list all of the people I need to thank, but I especially want to recognize:

- Sara Angus, Account Executive and former University of Washington lecturer, Lease Crutcher Lewis, commercial and custom residential general contractor (GC)
- Jeff Foushée, Founder, retired, Foushée and Associates, commercial GC
- Robert Guymer, Chief Operating Officer, Foushée and Associates, commercial GC
- Mark Hanson, Certified Public Accountant, Smith and Dekay, PS
- Bob Kendall, President, Star Rentals, Inc.

I would also like to thank Jane Holm, Suzanne Bailon-Schubert, and Sam Elliot for their research and contributions to the book and the instructor's manual. And last, and maybe most important, to all of those University of Washington construction management students who used drafts of this material in their construction cost accounting course as a trial run. Thank you and I hope you enjoyed the process.

There is a complete instructor's manual available on the eResource with answers to all of the review questions and many of the exercises. The instructor's manual also includes several case studies borrowed from the third edition of *Who Done It? 101 Case Studies in Construction Management* which will be re-published by Routledge in 2019 as *101 Case Studies in Construction Management*. This is an excellent economical companion book to many construction management topics.

If you have any questions about the material, or recommendations for changes for future editions, please feel free to contact the publisher, Routledge, or me direct at holmcon@aol.com. I hope you enjoy my connection of home office cost accounting operations to the construction jobsite.

Len Holm

Abbreviations

ABC	activity-based costing
ACWP	actual cost of work performed
AGC	Associated General Contractors of America
AHJ	authority having jurisdiction, often city building department
AIA	American Institute of Architects
APR	annual percentage rate
APY	annual percentage yield
AQWP	actual quantity of work performed
ARM	adjustable rate mortgage
B&O	business and occupation tax (also known as excise tax)
BCAC	budgeted cost at completion
BCWP	budgeted cost of work performed
BCWS	budgeted cost of work scheduled
BE	built environment
BIM	building information models or modeling
BOD	board of directors (also Board)
BQAC	budgeted quantity of work at completion
BQWP	budgeted quantity of work performed
BQWS	budgeted quantity of work scheduled
CAD	computer-aided design
Cap	capitalization (rate)
Carp	carpenter
CCA	construction change authorization
CCD	construction change directive
CDs	construction documents
CEO	chief executive officer

CFMA	Construction Financial Management Association
CFO	chief financial officer
CIP	cast-in-place (concrete)
CM	construction manager or management
CM/GC	construction manager/general contractor (also known as GC/CM)
CMBS	commercial mortgage backed securities
CO	change order
C of O	certificate of occupancy
COO	chief operations officer
COP	change order proposal
CPA	certified public accountant
CPFF	cost plus fixed fee
CPI	cost performance index
CPPF	cost plus percentage fee (similar to T&M)
CSI	Construction Specifications Institute
CV	cost variance
CY	cubic yard
DB or D-B	design-build (delivery method)
DBO	design, build, operate
DBOM	design, build, operate, maintain
DBOMT	design, build, operate, maintain, transfer
DCR	debt coverage ratio, also design change request
DD	design development (documents or phase)
DL	direct labor (estimate or cost area)
DM	direct material (estimate or cost area)
DRB	dispute resolution board
EA	each
ECC	Evergreen Construction Company (fictitious case study contractor)
EMR	experience modification rate
ENR	Engineering News Record (publication)
EPS	earnings per share
EV	earned value
EVM	earned value method
FE	field engineer
FED	Federal Reserve Bank
FHA	Federal Housing Administration
FICA	Federal Insurance Contributions Act (Social Security)
Fore	foreman
FUTA	Federal unemployment tax
FV	future value
G & A	general and administrative (overhead expenditures)
GC	general conditions, also general contractor
GDP	gross domestic product
GF	general foreman (similar to an assistant superintendent)
GMP	guaranteed maximum price, estimate or contract

GWB	gypsum wall board (also sheetrock or wall board or drywall)
GSF	gross square footage (building area but not necessarily leasable)
HO	home office
HOOH	home office overhead
HR	human resources, also hour
HVAC	heating, ventilation, and air conditioning (mechanical system or contractor)
IDS	income to debt service ratio
IPD	integrated project delivery
IRR	internal rate of return
IRS	Internal Revenue Service
ITB	instructions to bidders, also invitation to bid
JIT	just-in-time (material deliveries)
JV	joint venture
LDs	liquidated damages
LEED	Leadership in Energy and Environmental Design
LF	linear feet
LLC	limited liability company or corporation
LLP	limited liability partnership
LS	lump sum (cost estimate or contract)
LTV	loan to value ratio
MACRS	modified accelerated cost recovery system (depreciation)
MBF	thousand board feet
MC	mortgage constant ratio
MEP	mechanical, electrical, and plumbing, systems or contractors
MH	man-hour
Mil	million
MOCP	*Management of Construction Projects, a Constructor's Perspective* (textbook)
MXD	mixed-use development
NAHB	National Association of Home Builders
NAIOP	National Association of Industrial and Office Parks
NIC	not-in-contract, also not included
NLRB	National Labor Relations Board
NOI	net operating income
NSF	net square footage (building area available for lease)
NWR	Northwest Resorts, LLC (fictitious case study project owner)
O&M	operation and maintenance manual
OE	owner's equity, also operating engineer
OH	overhead
OH&P	overhead and profit (also known as fee)
OIC	officer-in-charge
OM	order-of-magnitude (cost estimate)
OSHA	Occupational Safety and Health Administration
PE	project engineer, also pay estimate
PM	project manager or management
PO	purchase order

PPP or P3	public-private-partnership (delivery method)
psi	pounds per square inch
PV	present value
QC	quality control
QTO	quantity take-off
Qty	quantity
Rebar	concrete reinforcement steel
Recap	cost recapitulation sheet (estimating)
REIT	real estate investment trust
RFI	request for information, or interpretation
RFP	request for proposal
RFQ	request for qualifications, also request for quotation
RII	Resorts International, Inc. (fictitious case study holding company)
ROA	return on assets
ROE	return on equity
ROI	return on investment
ROM	rough-order-of-magnitude (cost estimate)
ROR	rate of return
S&L	savings and loan institutions (banks)
SD	schematic design (documents or phase)
SF	square foot
SFW	square foot of wall
SIP	structural insulated panels
SOG	slab-on-grade (concrete)
SOV	schedule of values (estimate or pay request)
Spec or specs	specifications or speculation
SPI	schedule performance index
SPM	senior project manager
SUTA	state unemployment tax
SV	schedule variance
T&M	time and materials (contract or billing)
TCO	temporary certificate of occupancy
TI	tenant improvement
TN or Ton	tonnage (2,000 pounds)
TQM	total quality management
TVD	target value design
TVM	time value of money
Typ	typical
ULI	Urban Land Institute (development association and publications)
UMH	unit man-hours
UP	unit price
USGBC	United States Green Building Council
V	volume
VA	Veterans Affairs
VCT	vinyl composition tile

VE	value engineering
VP	vice president
VPR	variable payment rate
WA	Washington State
WBS	work breakdown structure
wks	weeks

1

Introduction

Financial management overview

Project engineers (PEs) and project managers (PMs) are not typically known as accountants. Their responsibilities as construction managers (CMs) generally revolve around the paperwork documentation performed at the jobsite to support superintendents and subcontractors in the construction of a building. And, although much of the financial management activities performed at the jobsite are not thought of as 'accounting' per se they do fall under the *cost accounting* umbrella, especially when we add *and financial management* in the title. In addition, everything that is done at the jobsite is connected with the operations of the construction company in the home office, and it is important that jobsite managers understand why they manage finances and how management of finances relates to the operation of the home office accounting department. This book is not meant to be a repeat of standard business-school accounting textbooks or accounting courses, there are many good ones available, but the focus here is on the financial operations of the construction jobsite team and how they interface with the accounting needs of the home office.

There are many detailed studies on cost accounting, including construction cost accounting. There are also many pure studies on project management, including construction project management, but the glue that ties these two topics together is a broader study of construction financial management. Financial management includes many CM topics, beyond just cost accounting, including:

- Estimating anticipated construction costs,
- Cost control,
- Cash flow projections and management,
- Processing invoices from subcontractors and suppliers,
- Processing pay requests to the project owner,
- Managing change orders,

- Financially closing out the construction project, and
- A variety of other advanced financial management topics such as activity-based costing, lean construction techniques, time value of money, taxes and audits, and the developer's pro forma.

Construction is a risky business. Construction failures are very high every year, especially with smaller start-up contractors. There are many statistics and metrics regarding how many contractors fail each year, and maybe just focusing on one year is too specific, but generally approximately 70% of the contractors in business on the first of any year will fail within seven years. Because construction is such a risky business, construction company owners or investors therefore expect a very high rate of return (ROR) on their investment. If all they could get on their out-of-pocket up-front investment in the company was 1–2%, then they would be better off putting their money in the bank where it is insured by the Federal Government and earning a guaranteed interest rate. In order to receive an acceptable ROR, contractors need to understand and manage their accounting and financial risks and responsibilities.

There are many causes or warning signs that a contractor might be in jeopardy of failing financially. These signs are important for not only the internal ownership of the company to be aware of, but also external strategic partners or stakeholders such as the contractor's bank and bonding companies, among others. The first and most common sign that financial difficulties may be boding is the lack of a good financial management plan or system. Some of the signs that a contractor may be suffering financial difficulty include:

- Inefficient financial management system,
- Borrowed on their credit line to the limit,
- Poor estimating processes and/or results,
- Poor project management systems and personnel,
- An adequate business plan is not in place,
- Internal and external communication problems, among others.

Often contractors think an increase in volume or total revenue will solve all of their financial problems. This is not necessarily the best solution. There are many reasons a contractor will choose to pursue construction work or feel they have the resources to do so. Some of the reasons they may choose to either bid or propose on a new construction project, or not to bid, include the following:

- Although contractors are not expected to provide the construction loan, as will be discussed later, they still must have a sufficient positive cash flow, especially early in the project;
- The contractor has sufficient bonding capacity which is especially important on a public bid project;
- Qualified and available employees are already on the payroll and ready to start a new project;
- The contractor has the necessary construction equipment, or immediate access to equipment;
- The home office overhead is staffed adequately with specialists to support the project team including estimators, schedulers, and cost accountants;
- They see a potential to make a reasonable fee;
- The contractor already has a positive history with the client, or is interested in a future relationship with the client, and/or they have a positive history with the architect or engineer, or are interested in a future relationship with those firms;

- This type of work is already a specialty of the contractor; and/or
- The contractor could use additional backlog.

The contractor's answers to all of these issues affecting their decision to pursue a project also impact the company's finances and approach to accounting both in the home office and at the jobsite.

Accounting purposes

There are several reasons a contractor should establish a formal cost accounting system, both at the home office and at the jobsite. Four of the more prominent ones include:

- Prepare financial statements for internal and external use;
- Process cash-in (revenue) and cash-out (expenditures);
- Prepare and file taxes as required by the Federal Government as well as some state and local jurisdictions;
- Manage the internal financial affairs of the company:
  - Are we making money?
  - Are we returning an adequate return on equity (ROE) to our investors?
  - Are we focusing on the proper type of work? Is our company operating in the proper market?
  - How are our people performing?

Consistency is important for cost accounting and financial management for contractors. Their financial reporting tools must be consistent from year to year, from project to project, and from month to month within each project. In order to report costs and projected profits consistently, contractors must have reliable financial management systems in place, particularly as it relates to cost control.

Cost control in construction is an important topic and is discussed in detail in several chapters of this book. It is important to distinguish between cost reporting and cost control, especially as it relates to jobsite cost accounting. The foremost question to ask is: Can the jobsite team really control costs or are they simply just reporting costs? And can they really 'control' the operations of the construction craftsmen in the field, or are they doing their best to 'manage' the process so that the craftsmen can achieve the estimate? Most construction management and cost accounting textbooks focus their cost control discussion simply on cost reporting. But if timely modifications and corrections are not made to the processes, the jobsite management team is not properly managing costs and cannot achieve the bottom-line fee, let alone improve it. To have an effective cost control system the construction project team must follow some basic rules:

- Cost reporting data has to be timely and accurate. If actual cost data was not input to the accounting system accurately, then the results will be of no value to the jobsite team.
- Eighty percent of the costs and risk on a project fall within 20% of the construction activities – this is known as the Pareto 80-20 rule. The jobsite team should focus on the riskiest activities. The 80-20 rule is expanded on throughout this discussion of financial management.

- The original estimate and schedule should be shared with the contractor's field supervision, including superintendents and foremen. In order for them to plan and implement the work they need to have been given the complete picture.

There are many bad examples of construction cost accounting, especially as it relates to cost control. We will be sharing some of these throughout the text. There are a few good ones sprinkled in as well. Here are a couple of examples to get you started:

Example One: This very large international contractor was constructing power plants in the 1980s that cost billions of dollars. There were 1,000 engineers in the office producing design for the 5,000 craftsmen in the field. Design changes and building code changes were occurring so often that it was easy to lose track of the original or current budget. Only two years out of college, this 24-year-old cost engineer was promoted to chief estimator on a $5 billion construction project, and although completely unqualified for the position, he gladly accepted it. At that time, the way the contractor forecasted next year's expenditures was to take last year's expenditures, regardless of where they were spent, and simply add 10% to it. It is no wonder these projects cost so much money. The proper way to create a cost forecast is explained later in this book.

Example Two: In April, six months after this large aerospace manufacturing facility was successfully completed, the owner of the drywall subcontractor called the general contactor's (GC) project manager and indicated that he was owed an extra $1 million, even though his contract had been closed out with proper lien releases. The subcontractor could not show why he was owed money other than it was tax time and he was preparing his books and realized the company lost money the previous year and this had been their largest project. He had not been using independent job numbers, let alone cost codes. Contract close-out and lien management are important financial responsibilities of the project team and will be covered in detail in this book.

Example Three: This superintendent wanted to teach the home office staff estimator a lesson on this out-of-town electronics facility construction project. He felt the estimator never included enough money for safety, so he charged $50,000 of concrete formwork to the safety cost code to make his point and completely blew the safety cost code out of the water. The proper way to create historical databases will be explained in several chapters in this book.

Introduction to the built environment

There are three major participants in all built environment projects. Chapter 3, 'Introduction to construction management,' will explain how these different firms might arrange themselves and contract with each other to deliver construction projects on different occasions. Every project has: clients, which will also be referred to as project owners or simply owners; designers, including architects and many different types of engineers; and construction managers and/or general contractors. There are also many consultants and subcontractors which will contract through or report to these three primary participants. One type of consultant that connects on many levels of cost accounting is the certified public accountant who may help all built environment (BE) participants prepare their books, especially for taxes and audits.

Construction types/industry sectors

Construction projects can be categorized into different sectors along the lines of type of building or how the facility is utilized. The following are examples of some broad construction industry sectors:

- Commercial, which includes retail, office, schools, fire stations, churches, and others. Entertainment and hospitality are subsets of commercial and includes movie theaters, bowling alleys, and restaurants.
- Residential including individual spec home and custom home construction. Residential also includes apartments, condominiums, senior housing, and hotels – especially if wood-framed as they include materials and methods similar to apartments.
- Heavy civil projects include bridges, roads, and utility projects.
- Industrial projects are very specialized including power plants, utility treatment plants, refineries, and others.
- Hybrids or mixed-use developments (MXD) will include two or three different uses, such as a downtown hi-rise hotel with underground parking, a restaurant, retail space, and luxury condominiums on the top floors.

Many construction companies are specialized in one of these industries, in that they classify themselves as commercial, residential, civil, or industrial contractors. Some contractors may overlap into a second industry or cover two different industries. For example:

- Commercial contractors may also build hi-rise residential apartments or hotels, as the materials and processes are similar to constructing an office building.
- Some commercial contractors may also have a high-end custom home division.
- Commercial contractors may also have a real estate development arm, and therefore contract work internally.
- Some speculative residential home builders also build custom homes and custom home builders may build spec homes.
- Larger home builders may also have real estate development arms and develop and design and build large tracts of homes on speculation for sale.
- Some larger residential contractors may also construct smaller commercial projects.

- Civil construction companies may also have a commercial division, but very few civil contractors would build residential projects.
- Specialty contractors are subcontractors which specialize in one or more facet of work, such as a roofing subcontractor or a landscape subcontractor. Specialty contractors work within all of these sectors and may specialize in one or more sector similar to general contractors.

Many of these sectors, and/or industry types, each have potentially different cost accounting and financial management applications. Those differences will be highlighted throughout this book.

Accounting managers

Chapter 3, 'Introduction to construction management,' will expand on the various BE firms and construction-company personnel. Roles of these people are defined throughout the book and clarify whether they are primarily located in the home office or at the jobsite. As it relates to cost accounting and financial management, the home office team includes the chief executive officer (CEO), the chief financial officer (CFO), and his or her supporting accounting department. Some of the accounting personnel may include positions such as the controller or comptroller, bookkeeper, payroll clerk, accounts payable clerk, accounts receivable clerk, and others. The jobsite team is headed up by the project manager and the site superintendent. Foremen and project engineers, who may also be known as the jobsite cost accountant or cost engineer, work for the superintendent and PM respectively. Contractual terms will have an impact as to who is located at the home office or the field and potentially what their titles might be on any particular project.

Accounting cycle

The cost accounting cycle includes many different steps and functions which overlap and interplay with phases of construction and cost control steps; all of these will be discussed in subsequent chapters. See Figure 1.1. Note that there are many dependencies between the home office accounting operation and the cost engineering conducted at the jobsite. Some of the cost accounting steps, which form the basis for the following chapters, include:

- Estimate database maintenance;
- Estimate preparation;
- Construction award and assignment of a job number by the accounting department;
- Buyout of subcontractors and estimate adjustment and assignment of cost codes;
- Development and implementation of foreman work packages;
- Expenditures realized in the field and reported to the home office and cost-coded;
- Pay request process including payments received from the client and payments made to subcontractors and suppliers;
- Incorporation of change orders into the project plan and the cost control system;
- Modifications to the cost control system if there are problems;
- Home office reporting by the PM including monthly fee forecasts;
- Financial close-out, including lien releases and retention; and
- As-built estimate and database update.

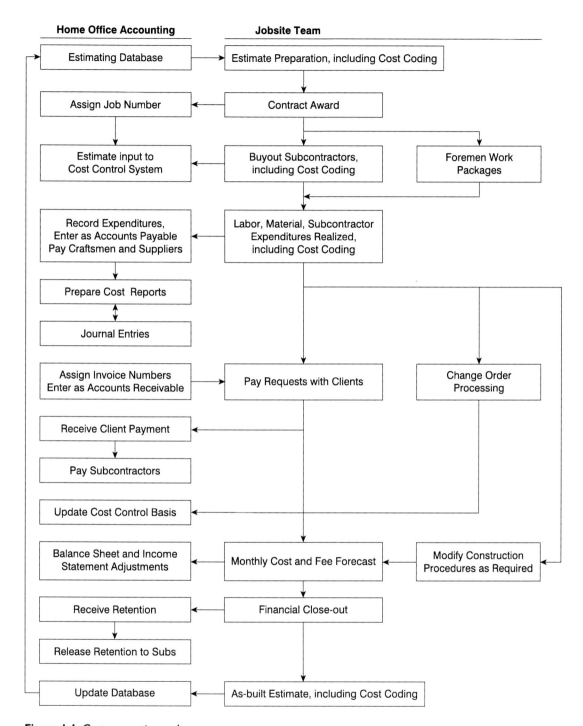

Figure 1.1 Cost accounting cycle

Organization of the book

Strict home office accounting and other financial management topics are blended in throughout this book, but the focus is primarily at the jobsite level. Introductory topics are presented in the beginning such as introductions to accounting, construction management, and estimating in Chapters 2, 3, and 4 respectively. Jobsite and home office overhead estimation and management are the subjects of Chapters 5 and 6. Accounting statements, primarily those from the home office, are explained in Chapter 7. Several chapters relating to many facets of cost control have been included, as well as topics of earned value analysis, activity-based costing, lean construction, site material management, and equipment management in Chapters 8 through 12. Additional aspects of cost accounting and financial management are in Chapters 13 through 16 including cash flow, pay requests, change orders, and financial close-out. The book concludes with advanced cost accounting and financial management topics on time value of money, taxes and audits, and the real estate development pro forma. A lot of abbreviations are used in construction and all of those introduced in this book are listed in the prelims. A glossary of terms is included at the end of the book.

Case study

Many of the examples, figures, and exercises used throughout this text will be based on one significant construction project which will be used as an integrated case study. This case study will thread through many of the chapters, connecting home office cost accounting and financial management topics with those of the jobsite project management team. This is a fictitious project and the project name and participants have been created just for use in this book. Any connection or similarity any portion of this case study has with a real construction project or real construction participants is coincidental. Some of the particulars of the project are listed in the following. Additional documentation such as a project organization chart, detailed estimate, and detailed schedule are available on the eResource.

- The name of the project is Olympic Hotel and Resort. The project is to be constructed on the Olympic Peninsula in Washington State. The project is also nicknamed the Olympic Hotel, or simply hotel for convenience in various parts of the book.
- The client is Resorts International, Inc., a national hotel chain which has considerable construction experience. They prefer negotiated open-book construction projects which utilize a guaranteed maximum price (GMP) contract. They always audit the contractor's books.
- The client forms an independent limited liability company for each hotel. In this case the ownership of the Olympic Hotel is Northwest Resorts, LLC (NWR).
- The hotel is located on a 10-acre site. The site has been prepared by a previous horizontal developer and civil construction contractor.
- The building is five stories tall. The first floor includes hotel reception, administration, an exercise room, and back of house areas such as laundry and housekeeping. In addition, the hotel has first-floor conference rooms and is targeting businesses and professional associations looking for a weekend or week-long retreat.
- The first-floor swimming pool design was not finalized at the time the GMP was executed but is anticipated to be change ordered into the project before completion. The first-floor

restaurant space is also currently shelled and is being marketed to outside restaurant chains for lease and management.

- Floors two through five house all of the hotel rooms, 30 on each floor, with a central hall. Rooms face either west towards the Pacific Ocean or east towards the Olympic Rain Forest National Park.
- There are two elevators.
- There is an additional above-grade, but separate two-story parking garage which will serve 150 cars which is scheduled to be built concurrent with the hotel.
- The first floor of the hotel is to be constructed of cast-in-place (CIP) concrete formed with rough boards which will provide it with a cedar plank architectural finish. The top four floors are wood-framed with cedar siding which blend in with the natural surroundings. Structural steel is used throughout for the main columns and girders.
- The architect, Gateway Design, has extensive hospitality and Pacific Northwest experience.
- The general contractor (GC) for the project was chosen on a competitive proposal basis. The client chose Evergreen Construction Company (ECC) whose main office is located in Seattle, Washington, approximately three hours from the project site. ECC has an extensive hotel resume.
- The contract NWR and ECC executed is an AIA A102 cost plus fixed fee with a GMP. The contract allows for all administrative personnel located on the jobsite to be cost reimbursable. It is because of this, and the distance from the home office, that ECC has decided to locate a full-time PM and a cost accountant at the jobsite. The jobsite cost accountant will also be referred to as a cost engineer. The cost accountant will be cost coded half time for accounting and other financial management functions and the other half time as a project engineer (PE). There is also one additional full-time PE and an administrative assistant on the site, and of course a salaried superintendent.
- The superintendent has three foremen who will be cost coding their time to the work. One is a carpenter by trade, the second a laborer, and the third is an ironworker.
- The contractor's GMP estimate is approximately $24.5 million, which includes a 5% fee. The jobsite general conditions estimate is developed from a detailed estimate and totals just less than 7% of the construction contract value.
- Corporate home office quality control and safety officers will visit the jobsite once weekly. Day to day quality and safety control is the responsibility of the project superintendent with the assistance of his foremen and the PE.
- The architect has prepared the drawings and specifications to allow the client to achieve a Leadership in Energy and Environmental Design (LEED) silver certification, if they so choose, which requires additional documentation and support from the construction team. The GC's PE is also LEED certified. The requirement to achieve LEED certification is not yet in the contract.

Review questions

1. Why do we as project managers and project engineers need to know about accounting at the a) home office and/or b) at the jobsite?
2. Which is the number one sign a contractor may be experiencing financial difficulties?

3. Of the construction company types discussed, which would most likely build:
    a. A golf course,
    b. The golf course clubhouse, or
    c. The home for the golf professional adjacent to the 18th tee box?

Exercises

1. Other than the earlier example, what mix of different construction uses (MXD) have you observed all in one structure?
2. Look ahead to Chapter 3's discussion of construction phases and Chapter 8's discussion of the cost cycle and compare those phases or steps with those of the accounting cycle in this chapter. There should be many overlaps.
3. What is the difference between cost 'control' and cost 'management'? You may have to look forward for this one as well.
4. Other than the reasons given earlier, why might a contractor choose to bid a project?
5. Other than the signs given earlier, how might you notice a contractor, or subcontractor, is suffering financial difficulties?
6. Assume every student in your class starts a construction company next June. How many of them will declare bankruptcy within seven years of graduation? Hopefully not you!
7. Without looking at the organization chart on the eResource, draw one for the case study project from the information given earlier. You may want to thumb through the rest of the text for additional personnel and company names. After you are done, check it with the website. Did we forget anybody?
8. If you have already worked for a construction company, what subset of the construction industry did it specialize in?

2

Accounting methods

Introduction

The last chapter introduced the reader to the book, the built environment (BE), our case study, general accounting basics, as well as the jobsite in lieu of a home office accounting and financial management approach. This chapter provides an additional introduction to accounting basics, including why construction differs from other industries and warrants its own unique approach to accounting. The construction industry is one of the largest industries in the United States, contributing approximately one trillion dollars ($1,000,000,000,000) annually to the gross domestic product (GDP). That is a lot of zeros! This chapter examines construction firm ownership structures and how those structures affect the four different accounting methods. Accounting is performed for both external and internal stakeholders and one use of accounting is to make sure the contractor's assets are protected and properly managed.

Uniqueness of the construction industry as it relates to accounting

There are many differences between the construction industry and other mass-production industries such as automobile manufacturing or retail sales or food service. Because of these differences, contractors must take a different approach to accounting and other forms of financial management. Some of those differences are listed here:

- Construction is job-based or project-oriented. Each construction project is given a separate job number from the accounting department immediately after the contractor is notified of award from the client. Each project is managed independently by project teams, including the project manager (PM) and the superintendent, and is considered a separate profit source.
- Other industries are product-based, not project-based, and provide repeat products, such as a pizza restaurant producing ten different varieties of pizzas and on any given Friday night

they might prepare 50 pizzas of each of those different types. It is very unlikely they know how much each of those pizzas cost.

- In construction each project is unique. Most construction companies provide variable products. Even speculative home builders, or track developers, have variable products. They may build 200 homes in a tract development with five different floor plans, but they provide several different facades and colors and interior finishes such that buyers feel their home is unique. In the next tract they may build 150 homes with seven different plans, all of them unique from the first tract choices. And all of these are on separate building lots.
- Each construction project utilizes thousands of variable parts and materials.
- Contractors retain very little inventory or unused materials. Speculative home builders may have a few homes finished ahead of their sales, but they try not to become over-extended.
- Construction projects are decentralized. Each project is built on a different site in a different location – they are not all near the contractor's home office.
- Each project utilizes different construction equipment and is built with a different workforce.
- The construction industry experiences irregular cash flow situations. Cash flow is defined as revenue coming in to the company in the form of monthly client payments, and expenditures are cash leaving the company in the form of labor and material and subcontractor payments.
- Most construction projects are subject to retention in that the client retains 5 or 10% of each month's pay request to make sure the contractor finishes the project. This would be akin to you holding $2,000 back from a car dealer for a couple of months to make sure your car operates as they purported. Would they allow that?
- When you buy a hamburger, the business transaction is complete when you walk away from the cash register. This is a short-term contract. Construction projects have long-term contracts, which typically last six months to a year or two, and project/product warrantees that last a year, or more, longer than that.
- In construction, builders execute a contract and have settled on the sale or total price of the project, but they haven't spent any money yet and will not know what the project will cost until construction is actually complete.
- When you buy a shirt at the clothing store they expect you to pay them 100% at the time of purchase. Speculative home construction is similar. But most facets of construction rely on partial monthly bills to the client and monthly payments spread out over the duration of the contract.
- General contractors (GCs) rely on specialty contractors or subcontractors to perform 80–90% of the work. Construction management companies may subcontract out close to 100% of the work. Imagine buying French fries and each of the 30 fries in the basket came from a different restaurant that employed a different cook who utilized a different deep fryer. Then the burger joint you were purchasing the fries from grouped them all together and sold them to you in one order.
- Each construction project will have a different mix of subcontractors and craftsmen. Very small residential remodeling contractors might use the same subcontractors and the same craftsmen on each project – but again, each of those projects would be for a different scope on a unique site and for different clients.

The bottom line is *Construction is Unique*. There will never be two identical construction projects with the same client, architect, GC, subcontractors, material suppliers, craftsmen, equipment rental, building sites, weather conditions, built on the same dates of the same year, and on and on. Even the following example was unique.

> Example One: This experienced owner's representative managed six elementary schools built for the same school district with the same architect and the same basic design (there were small differences) and built by the same general contractor. This was as close to repeat construction product that he had ever witnessed in construction. But each of these schools were built in different years, on different construction sites, utilizing a similar but different mix of subcontractors and craftsmen, and each school had a different school principal who had his or her own unique priorities.

Construction accounting

Accounting can be performed for construction companies in a variety of fashions as well. Some of the types of accounting services contractors utilize include:

- Some contractors employ outside accounting firms which employ certified public accountants (CPAs) to do all bookkeeping with little to no accounting support performed internally by the contractor. This would only be the case with very small or start-up contractors.
- Many contractors have an internal home office accounting department which is staffed with bookkeepers, accounts payable clerk, accounts receivable clerk, payroll clerk, controller or comptroller, and maybe someone responsible for accounting of the construction equipment and maintenance of the equipment ledger. These departments are usually headed up by a chief financial officer (CFO), although smaller firms may have all of these duties performed by one accountant who reports to the chief executive officer (CEO) who then relies on the service of an outside CPA for tax and other external reports.
- Jobsite accounting is usually only performed on very large construction projects or those cost-reimbursable open-book projects which will only pay for accounting support when it is located at the jobsite. In addition to accounting, the jobsite team has several other financial management responsibilities such as cost control, pay requests, and financial close-out, and others that are discussed throughout this book. The jobsite financial management team includes the project manager (PM), superintendent, project engineer (PE), and potentially a jobsite cost accountant or cost engineer. Look ahead to the sole-source organization chart depicted in Figure 3.8. This is the scenario that has been adopted to use for the case study and examples throughout this book. The focus will be on project-specific site-performed accounting and financial management functions, but all the while relate them to home office oversight.

Construction is a risky business for a variety of reasons as stated here and later. It is up to each jobsite PM and superintendent to manage those risks on behalf of the company they are representing. The jobsite team and the home office utilize a variety of financial resources. Some of these are originated in the home office, often with input from the jobsite team, and other reports are generated at the jobsite. One goal of this book is for the reader to develop a general understanding of these financial reports and processes and their importance, not necessarily that you will be able to create them yourself. Managing jobsite costs and returning a fair profit for the firm is a major goal of the jobsite team. Management of cash flow is an important construction management tool; just as a hammer is an important tool for a carpenter. Managing cash is first done at the jobsite level and then at the corporate level. The lack of positive cash flow is a big source of contractor failures. Cash flow is an important topic explored throughout this discussion.

A lot of built environment participants use different terms for different purposes. Many college textbooks, including construction management and accounting textbooks, also vary to some degree, especially as they relate to job titles. Some of the different terms and positions as they relate to construction cost accounting and financial management include:

- The boss or owner of a construction company may be the CEO or President, but there are a variety of other officers that the jobsite team may report to including officer-in-charge (OIC), chief operations officer (COO), senior project manager, and others which are also described in the next chapter, 'Introduction to construction management.'
- Home office financial manager and/or general manager include the CFO and CEO.
- Home office staff support members include the accounting team listed earlier as well as staff estimators, staff schedulers, specialty superintendents, quality control officers, safety control officers, human resource managers, marketing directors, and others.
- The jobsite team or management team includes the superintendent, PM, PE, and jobsite accountant or cost engineer.

Construction firm ownership structures

General contractors, subcontractors, suppliers, consultants, and other members of the built environment industry, including project owners and architects, can be either individuals or companies. The most common forms of business organizations are sole proprietorships, partnerships, corporations, limited liability companies, and joint ventures. Every company involved in construction including the client, GC, subcontractors, and design firms have an 'owner' or group of owners or equity partners. The form of a business organization has an effect on how the contractor operates its business in the form of decision-making, responsibilities, profit taking, risks and liabilities, and tax obligations. Because of these variations, the contractor's legal organizational structure impacts its approach to cost accounting and the financial management of the construction firm. This book will endeavor to distinguish the project owner from the construction firm ownership in our examples. The balance of this section is referring to the ownership of construction companies, not necessarily the client's ownership structure.

Sole proprietorships are the most common business form operating in the built environment. Proprietorships are popular because they are easy to start and maintain a business that is owned and managed by one individual, the proprietor. Therefore, there is not a distinction between the business and its owner who is entitled to all of the profits but is also fully responsible for the business's actions. Requirements for forming a proprietorship change depending on the state having jurisdiction. All permits and business licenses required to operate a construction company need to be obtained by the proprietor. For instance, a single carpenter can operate as a small deck-building business, but he or she would still need to obtain insurance, bonding, and a business license, as dictated by the jurisdiction where he or she operates. It is easy, but not necessarily advisable, for the proprietor to mix business income and expenses with that from other personal sources. Mixing of business and pleasure may have unfortunate tax implications.

An additional advantage of operating as a sole proprietorship includes flexibility. The proprietor makes all decisions regarding where he or she wants the business to go, growth and expansion opportunities, and speed of decisions and responses. All of the sole-proprietor's decision-making is made by a single person. Conversely the business owner can decide at any time to restrict or slow business, terminate the business, or sell it. The disadvantages of operating a proprietorship include its unlimited personal liability, its limitations in raising money to grow and/or operate the business, and its heavy decision-making burden, which falls solely on the proprietor.

Partnerships are businesses where two or more individuals share ownership in the company. The partners' initial contributions to the company could be in the form of cash or assets, such as construction equipment, and additional contributions of time and/or assets potentially throughout business operations. Similar to a proprietorship, a partnership is closely associated to its investors. Profits are passed 100% through to the partners, similar to a sole proprietorship, and are taxed with other personal gains. However, a partnership is a distinct entity that can independently own property, hire employees, and undertake (or be the subject of) legal action. The partners do not have to hold equal 50-50 shares. One partner may have contributed 75% of the original company equity whereas another contributed 25%. The two would then distribute profits along these same ratios.

Partners can be of two types: *general* and *limited partners*. General partners share responsibilities, and profits and losses of the business. If not specified otherwise, general partners will distribute those profits, liabilities, and management duties equally. Limited partners only contribute assets and do not participate in daily management or operation of the business. In return for their limited involvement, limited partners do not participate in corporate liability beyond their initial contribution. Adding limited partners is a way for a partnership to raise capital while assuring personal immunity to the investor and complete control for the general partner. Every partnership must list at least one individual as a general partner.

Difficulties associated with partnerships arise during the transfer of one partner's share if he or she dies, has a disagreement with the other partners and desires to dissolve the business relationship, or if a partner divorces and their spouse wants to retain a one half share in the construction company.

Two types of corporations exist, type C and type S. *Type C corporations*, or simply corporations, are legal entities owned by shareholders. Smaller contractors that originated as a proprietorship or a partnership may become a corporation by 'incorporating.' Forming a business as a corporation is a way to legally separate the equity ownership of the construction company from the management and operations of the company, which limits the liability to the shareholders to only their initial contribution. Shareholders appoint a board of directors, which selects and oversees business operations led by officers, including a CEO, COO, and a CFO. Equity owners in a corporation are often officers and active participants in the operation of the construction company and in return also receive a salary.

One additional reason a contractor may decide to incorporate is that the corporation will generally pay a lower tax rate than would a proprietorship or partnership, especially at very high income levels. Different from proprietorships and partnerships, corporations file taxes distinct from their shareholders. Profits of a corporation are taxed at a corporate tax rate, which is often lower than a personal income tax rate. If the management decides to distribute part of the profits among the shareholders, these profit distributions are known as *dividends* and are subject to a second Federal income tax at the individual level. Evergreen Construction Company (ECC) is the general contractor utilized in the Olympic Hotel and Resort case study in this book and they are organized as a type C corporation.

Type S corporations are smaller closely held corporations with limited shareholders. These are often new or start-up construction companies which will later transform into C corporations. One advantage of the S corporation is it avoids double taxation on dividends and the company is taxed at individual tax rates, which are lower at reduced income levels.

Limited liability companies or *corporations (LLCs)* combine features of corporations and partnerships. Owners of a LLC are sometimes referred to as members. Similar to a corporation, a LLC would limit personal liability from ownership. Similar to a partnership, a LLC would avoid double taxation by passing through all of its profits and losses to its members. Many real estate development projects are organized as separate individual LLCs protecting the development company and equity partners from failure on any one individual project. Real estate development companies and their financial pro forma is the subject of the final chapter of this book.

Joint ventures (JVs) are business entities that resemble general partnerships. The main characteristic of a JV is its temporary scope as they are created for only a limited period of time or for a single construction project. The main reasons for using a JV on a construction project are to pool financial resources and/or physical and human resources, and for spreading risk. For example, many large infrastructure projects are built by JVs of construction companies that pool their personnel, equipment, and bonding capacities while spreading the risk of failure. Chapter 18 discusses some of the tax differences between all of these different types of construction company ownership structures.

Construction accounting methods

As a member of the jobsite management team you are not necessarily going to be involved with the accounting processes and methods utilized by the home office accounting department and the CFO, but you should know some of the major issues. There are basically four different types of accounting processes available to construction contractors which include *cash, accrual, percentage of completion,* and *completed contract.* Each of these four methods will be described in this section in turn. They are differentiated in the processing and timing of revenue and expenditure recognition, or when they 'account' for money. The cash and accrual accounting methods are more applicable for short-term construction projects, those lasting less than one year, and percent complete and completed accounting methods are best suited for construction companies with projects lasting longer than one or two years. Accounts receivable includes the steps necessary to invoice a client for work performed and receipt of revenue in the form of payment. Accounts payable involves commitments made to process expenditures for labor, materials, and subcontracts for construction work performed.

Cash

The cash accounting method is usually reserved for smaller contractors, such as those with less than $5 million in annual company volume. Approximately 80% of construction companies employ ten or fewer employees and use the cash method. Contractors don't actually work with hard currency in construction, but the term 'cash' generally means a physical exchange of checks or electronic debit transfers from a bank account. The cash method is very simple. Contractors typically use a fiscal year that runs from January 1 to December 31. If a check is received from a client and deposited into the bank, it is reported as revenue. If a check is written to a subcontractor, it is reported as expenditure. The checks themselves do not have to clear the bank for either to be reported. This system does not consider work performed and not yet invoiced, nor a pay request invoiced to a client but not yet received, nor invoices received from subcontractors but not yet paid. The cash accounting method reflects very uneven cash flows. This uneven cash flow may cause problems for auditors, CPAs, the Internal Revenue Service (IRS), and equity partners who all look for consistency in financial reporting.

At the end of the year the contractor has some tax-planning flexibility in either depositing a client's check in December or holding it until January to postpone tax liabilities. The contractor may also process invoices from subcontractors early in December, although they may not be due until January, therefore moving costs or expenditures forward so that they may be deducted from this year's taxes. The cash method is also popular with designers and consultants such as architects, engineers, inspectors, and agency construction managers. Chapter 3 will describe the roles of many of these contributors to the built environment.

Accrual

The accrual accounting method reports revenues when they are billed or earned, and become accounts receivable, but not necessarily received. Retention held by the client is not reflected as income until it is received after project close-out. Expenses are recorded when costs are incurred and subcontractor invoices are invoiced and become accounts payable. The accrual method provides a more consistent financial reporting trail than the cash method. The accrual method is generally limited to smaller contractors, similar to the cash method, which have annual volumes of $5 million or less.

Percentage of completion

The percentage completion method is recommended for commercial contractors which have a volume greater than $10 million and/or have long-term contracts which may last longer than two years and overlap from one fiscal year into another. It is considered a more accurate and consistent indicator for recognizing income. This accounting method simply reports the percentage a construction company is complete at the end of the fiscal year with all of their construction projects. This is accomplished by reporting the percent of estimated costs incurred, and the proportional share of profit earned. In the case of a construction company with more than one project this is the summary of all of its project percentages completed. This is similar to invoicing against a pre-established schedule of values as will be discussed later. In the percentage of completion accounting method, the retention is also recorded at the completion of the project, similar to accrual. Larger architects and engineers may also utilize percentage completion financial reporting. The percentage of completion method will be the basis for the case study in this book.

Completed contract

The completed contract accounting method is even more irregular than the cash basis. In this case, contractors do not recognize any revenue or expenditures until the project has been completed and they are paid. This method may be used for small general contractors, such as those which do remodeling projects and speculative home builders, which are paid in full upon completion. Residential tract builders are contractors which build and sell homes on a speculative basis (also known as spec home builders). They are essentially real estate developers. In the completed contracting accounting method, jobsite administration expenses and home office overhead are proportioned to projects and expenses are reported when revenues are received, at project completion, with respect to tax implications. Because a home builder may sell ten houses in one year and only two in another, based on timing of construction starts or changes in the economy, revenues may change dramatically from one year to the next with the completed contract accounting method.

Internal accounting controls

There are many purposes of construction cost accounting as was briefly introduced in Chapter 1. One major external reason of accounting for revenues and expenditures includes reporting net income and paying taxes. There are also a variety of internal reasons for accounting which assists with management of the financial affairs of the company. Those include measurement of profits and their net effect on investor's equity; known as return on equity (ROE). As discussed, construction products are not repetitive, rather they are variable. Internal financial reports allow the contractor to perform self-diagnosis on industry specialization. Are we operating in the proper segment or focus area of the industry? Is one geographical area better for the company and our employees than another? Is our focus on the proper type and size of work?

Internal accounting, or audits, also allows contractors to measure the performance of their most important asset, their people. Construction is project based, and each project is staffed with a project manager and superintendent who need to return a fair profit to the company. One additional reason that contractors perform internal accounting management is to ensure that jobsite and home office teams are properly managing the corporation's money. The superintendent has several financial management responsibilities on the jobsite, many of which will be discussed throughout this book. Protection of construction materials from weather and external theft, and jobsite safety and security, are under the superintendent's control. Although not always considered as financial concerns, if not addressed properly these aspects will have a negative effect on project costs. Internal financial theft is unusual, but it can happen. There are several ways a contractor can ensure that there are not any internal financial thefts. Some of these include:

• Separate duties assigned to separate individuals such that no one person has complete financial control.

- Keep a good paper trail, even though most of the paper trail today is produced on the computer.
- Only appropriate persons should be issuing subcontracts and purchase orders and signing approval on invoices.
- Although contractors typically do not have significant inventory or assets, what assets they have, such as construction equipment, must be tracked and accounted for.
- Routine accounting operations include an internal financial system of checks and balances. Internal audits are also discussed in Chapter 18.
- The contractor should limit access to its financial reports to upper-management personnel only.
- Multiple approval signatures shall be required on as many documents as possible, especially material and subcontractor invoices.
- Even large contractors with substantial accounting departments will employ outside accounting assistance from CPAs to audit their own books for internal reporting as well as prepare tax reports.

Summary

Construction is the foundation of the built environment industry and contributes significantly to the GDP. Construction is unique from all other industries and therefore warrants a unique approach to cost accounting. Smaller contractors may employ outside accounting services which include certified public accountants. Most contractors employ a combination of internal accounting personnel but still retain the services of a certified public accountant for audits and taxes. Depending upon the job size and location and contract terms, cost accounting and cost engineering may be performed either at the jobsite or in the home office.

Contractors can be organized in a variety of different business formats including proprietorship, partnerships which may include general and limited partners, C and S type corporations, limited liability companies, and joint ventures. Most of the larger commercial general contractors are organized or incorporated as type C corporations. There are four different methods utilized for construction cost accounting including cash, accrual, percentage of completion, and completed contract. The choice of which one to use depends upon a contractor's size, organizational structure, and area of specialty. There is not one best legal structure or accounting system for all contractors; they all have their advantages and disadvantages under different circumstances. There are many reasons to perform cost accounting including measuring and monitoring internal financial success.

Review questions

1. What is the difference between a C corporation and an S corporation?
2. What is the difference between a CEO and a CFO?
3. Why would a construction company want to perform accounting operations out of the home office?
4. Why would a construction company want to perform accounting operations at the jobsite?
5. What is a CPA? Is this a person or a company?

6. Why would a contractor employ the services of an outside CPA versus an internal bookkeeper?
7. Who might the PM report to other than a CEO?
8. What type of construction firm ownership structures are subject to double taxation?
9. What is the difference between a general and a limited partner?
10. List at least two other staff positions in the home office beyond those provided in this chapter.

Exercises

1. Why do you feel a speculative home builder does not carry 50 completed homes in inventory?
2. Other than the examples given earlier, what other industries are drastically different than construction?
3. What industries are similar to construction?
4. List another reason that the construction industry is different than other industries.
5. If you already work for a construction company, which of the ownership structures discussed earlier applies to your firm?
6. Not discussed in this text, but what items on a construction project might be vulnerable to external theft?
7. Construction CFOs are officers and often part-owners of the company. Why do you think that is?
8. This author's construction consulting business is a sole proprietorship and utilizes the cash accounting method. Why do you think he made those choices?
9. Compare the four primary methods of construction accounting discussed in this chapter and compute the reported revenue, expenses, and income or profit for each, for a contractor which has only one project for the whole year as reflected in the following. Assume January 1 through December 31 fiscal periods. Ignore any retention or tax considerations for this exercise. The reporting date is 12/31/2019.
   - Contract volume of $1 million.
   - Included in the contract value is a 5% fee.
   - $500,000 was invoiced and received as of 11/30/2019.
   - All invoice values include a proportional share of fee.
   - An additional $100,000 is invoiced as of 12/31/2019, due 1/10/2020.
   - $450,000 has been paid by the GC in labor, material, and subcontractor expenses through 12/15/2019.
   - $90,000 was invoiced by subcontractors and suppliers for month ending 12/31/2019, payable ten days after receipt of payment from the client. That is factored into this month's $100,000 pay request.
   - The balance of the construction costs will be paid in full by 3/31/2020 by the client to the GC and the GC to its subcontractors and suppliers and craftsmen.
   - Theoretically the contractor's expenses will exactly equal its estimated costs, but it never happens this way.
10. Assume the contractor in Exercise 9 did not have any additional business in 2020. What revenues, expenses, and profits would be reported for that fiscal year for each of the four different accounting methods?
11. The construction industry today is ranked _____ among other GDP industries.

3

Introduction to construction management

Introduction

This chapter has been included along with other *Cost Accounting and Financial Management* topics to provide a brief overview and introduction to current construction management (CM) processes. It is important to introduce the reader to a few additional CM terms and processes before diving deeper into the accounting of construction. Neither this chapter on CM nor the next introductory chapter on estimating are meant to be stand-alone treaties on those very important topics. Each of those deserves their own book or separate class on just those areas. Our focus here is on the relationship between project management and estimating, and other CM topics, as it relates to jobsite cost accounting and financial management. There are many good books on project management and estimating including *Management of Construction Projects, A Constructor's Perspective* (*MOCP*) by J. Schaufelberger and L. Holm (2017), and the reader may want to look to that book for a more in-depth coverage. Much of the material in this chapter has relied upon that resource.

The focus of construction management here is on the role of the project manager (PM) as an individual. Many concepts and terms from one (CM or PM) apply to the other. The CM and the PM are both builders, but so are many other members of the built environment (BE) team. The term 'builder' is too generic; everyone on the project, including the owner, architect, general contractor (GC), and the craftsmen contribute to the building process and each is a 'builder' per se. The owner-architect-GC-craftsmen were in the beginning all one and the same; they were the builder. The qualities of the ancient builders included creativity, resourcefulness, intuition, problem solving, among others. They just 'knew' what it took to get the job done. Later the owner and the builder, or master-builder, split into two separate functions. It was the owner who provided the need and the builder who satisfied those needs, in this case, with a building or a home. At that time the master-builder was both the designer and the builder, similar to the current design-build (DB) delivery method described later and shown in Figure 3.4.

The next split was between the builder and the designer. The builder would evolve into today's construction management or general contractor companies. The design element would also later be split into architecture and engineering. Each of those also has many sub-divisions today and has been listed later. There are also many additional levels of specializations within the GC's organization as discussed later.

The project owner, sometimes also referred to as the client, and the lead designer or architect, and the general contractor or construction manager are the three major participants in any organization or delivery method utilized in the BE as will be discussed in the next section. There are many other BE participants including specialty engineers and consultants such as waterproofing or elevator consultants; the list is quite extensive. Many of these are listed here:

- Structural engineer
- Civil engineer
- Mechanical engineer
- Electrical engineer
- Kitchen consultant
- Landscape architect
- Interior designer

- Elevator consultant
- Accountant
- Bank
- City inspectors
- Third-party inspectors
- Commissioning agent
- Envelope consultant/waterproofing

This chapter will describe the delivery and procurement methods the client chooses to use as well as pricing and contracting options which the contractor must incorporate. Different types of general construction organizations are also introduced as well as the role of project management and project managers and the risks they must evaluate when considering a potential construction project.

Delivery and procurement methods

The three major companies and/or individuals which are the primary responsible parties in any construction project include the project owner or client, the designer (architect or engineer), and the general contractor. The relationships among these participants are defined by the delivery method used for the project. The choice of delivery method is the owner's, but it has an impact on the responsibilities of the contractor's jobsite team, especially their financial responsibilities. The project owner selects a delivery method based on their experience and the risk they are willing to absorb or pass to either the GC or the design team. This section examines the four most common delivery methods. Each of these different arrangements has an effect on how the GC will conduct its jobsite project management operations, including cost accounting. Basic differences in bid and negotiation procurement processes are also introduced along with the four major design phases which coincide with various delivery and procurement options.

Traditional general contractor delivery

The most common method or project delivery is the traditional method and is represented in the simple Figure 3.1 organization chart. The owner has separate contracts with both the designer

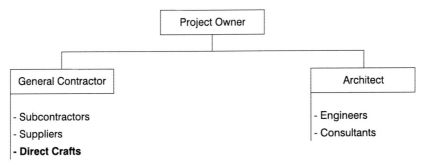

Figure 3.1 Traditional organization chart

and the general contractor. There is no contractual relationship between the designer and the GC. Typically, the design is completed before the contractor is hired in this delivery method. The GC's project manager takes the point to obtain the project drawings and specifications, develop a cost estimate and construction schedule, establish a cost control system to manage construction financial activities, and manage the construction site office. A traditional GC employs a mix of subcontractors and direct craftsmen. The hotel case study and exercises used throughout this book are based upon the traditional GC delivery method.

Construction management delivery

There are two basic construction management delivery methods. One is the agency CM and the other is the at-risk CM. The agency CM does not employ any subcontractors or direct craftsmen and the pure at-risk CM employs all subcontractors and no direct craftsmen; different than the mix employed by the traditional GC as described earlier.

The project owner has three separate contracts (one with the designer, one with the general contractor, and one with the construction manager) in the *agency construction management* delivery method as shown in Figure 3.2. The CM acts as the owner's agent and coordinates design and construction issues with the designer and the GC. The CM usually is the first contract awarded,

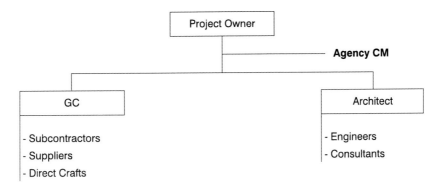

Figure 3.2 Agency construction management organization chart

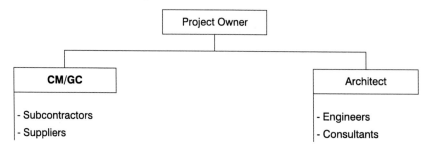

Figure 3.3 Construction management at-risk organization chart

and he or she is involved in hiring both the designer and the GC. In this delivery method the GC usually is not hired until the design is completed. The GC's project manager has similar responsibilities to those listed for the traditional delivery method. The primary difference is that the GC's PM interfaces with the agency CM instead of the owner, as is the case in the traditional method. The agency CM is sometimes referred to as the owner's representative. Since the construction costs do not run through the agency CM's accounting books, this version of the CM does not have any financial risk.

In the *construction manager-at-risk* delivery method, the owner has two contracts (one with the designer and one with the construction manager) as illustrated in Figure 3.3. This delivery method is also known as the *construction manager/general contractor* (CM/GC) delivery method. In this case, the designer usually is hired first, but the CM/GC is also contracted early in the design development to perform a variety of preconstruction services, such as cost estimating, constructability analysis, and value engineering studies. Once the design is completed, the CM/GC constructs the project. In some cases, construction may be initiated before the entire design is completed. This is known as fast-track or phased construction. The CM/GC's project manager interfaces with the designer and manages the execution of preconstruction tasks and assumes similar responsibilities to those in the traditional method once construction starts.

Design-build delivery

The project owner has a single contract with a design-build (DB) contractor for both the design and construction of the project in the DB delivery method, as diagramed in Figure 3.4. The DB contractor may have a design capability within its own organization, may choose to enter into a joint venture with a design firm, or may hire a design firm to develop the design. Construction may be initiated early in the design process using fast-track procedures or may wait until the design is completed. In this delivery method, the GC's project manager interfaces with the owner and manages both the design and construction. This method is very similar to the ancient master-builder concept discussed earlier in this chapter. There are other versions of DB which extends the contractor's involvement further into the life-cycle of the building, including:

- DBO – Design, build, and operate,
- DBOM – Design, build, operate, and maintain, and
- DBOMT – Design, build, operate, maintain, and transfer.

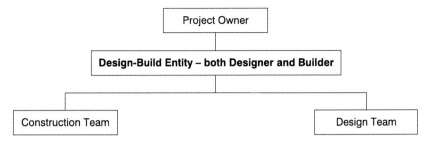

Figure 3.4 Design-build organization chart

There are a variety of other delivery methods, some of which are hybrids of these four. The *integrated project delivery* (IPD) is a relatively new concept where all three prime parties sign the same agreement and share in the same risks. A sample IPD organization chart is included later in Chapter 11, 'Lean construction techniques.' In the five primes or *multiple-prime delivery* method, the project owner contracts with several general contractors or major subcontractors at the same time. This often includes a site work contractor, building shell and core contractor, tenant-improvement contractor, mechanical contractor, electrical contractor, and/or others. The multiple-prime delivery method usually requires a construction-experienced project owner to act as their own construction manager. Another alternate delivery process is known as *public-private-partnership* (PPP or P3). This process combines the resources of a private entity, such as a real estate developer which may own a piece of property, with the needs of a public client, such as a university looking to build a research laboratory. The private developer will build and operate the facility for benefit of the university which will sign a long-term lease guaranteeing the developer's pro forma will work out. The public client does not need to go through the sometimes complicated and often litigious lump sum bid process with the P3 delivery method.

Procurement

Project owners solicit or procure both construction and design team members utilizing either bid or negotiated procedures. Public project owners are often required by law to use an open bidding process to allow all contractors an even chance of successfully landing new work. Private owners can use whatever procurement method they choose, bid or negotiate, but often solicit contractors they have had good experience with in the past, and may ask a select few or even only one contractor to submit a bid or negotiated proposal. Both private and public owners may use a prequalification process to develop a short list of contractors to either bid or negotiate after they have submitted a list of qualifications which suit that specific project. Prequalification helps reduce the client's risk of an unqualified contractor submitting a bid which is too low for them to successfully complete the project. Major subcontractors and suppliers can also be prequalified and vetted which also helps reduce the risk for the bidding general contractors. There are risks and advantages associated with either bidding or negotiating for the owner, architect, and contractor. This book will utilize a contractor chosen for a private project on a negotiated basis as an example case study and the basis for many of the exercises.

The term 'book' or 'books' is referred to throughout this discussion on financial management. This is in the context of bookkeeping or accounting books. Essentially this includes the contractor's original estimate, all of the accounting statements and reports and actual cost records collected throughout the course of construction, among others. Bid projects are closed-book in that the client cannot look at the contractor's actual detailed estimates or costs. There is more to focus on with respect to a study on jobsite cost accounting and financial management when considering an open-book project that provides additional accounting backup for change orders, pay requests, and financial close-out audits.

Design phases

All elements of any given project are not designed at the same time; different elements are started earlier, and some proceed faster than others and include additional levels of detail. There are five major design phases that most built environment projects experience. In some cases, programming and conceptual design may be combined into one phase for a total of four phases. These design phases influence all of the other introductory construction management topics discussed in this chapter including delivery, procurement, contracting, and pricing methods differently with respect to when the contractor estimates the project and when and how the contractor is chosen and begins their cost accounting processes. The design phases include:

- Programming,
- Conceptual design,
- Schematic design (SD),
- Design development (DD), and
- Construction documents (CDs).

Additional design definition and detail is added at each of these levels which allow the contractor to develop more accurate estimates and reduce contingencies. Ideally this additional detail also reduces the client's risks. Budget estimates and negotiation with a construction manager/general contractor is more common early in the design process, whereas the traditional delivery method and bid procurement process matches up best at the completion of the last phase, construction documents.

Pricing

There are several methods for pricing contracts used in the construction industry but four are the most prevalent and others are hybrids of these four. The choice of which to use on any particular project is made by the owner after analyzing the risk associated with the project and deciding how much of the risk to assume and how much to pass on to the contractor. Contractors want compensation for risk they assume and usually do so with increased fees or estimating contingencies. The most common pricing methods include:

- Lump sum (LS),
- Unit price,
- Cost plus fixed fee or cost plus percentage fee, and
- Cost-plus with a guaranteed maximum price (GMP).

Lump sum, or stipulated sum, contracts are awarded on the basis of a single lump sum estimate for a specified scope of work. Unit-price contracts are utilized for heavy-civil projects when the exact quantities of work cannot be defined. The designer estimates the quantities of work, and the contractor submits unit prices for each work item. The actual installed quantities required are multiplied by the bid unit prices to determine the final contract price but only after project completion. Cost-plus contracts are used when the complete scope of work cannot be defined. All of the contractor's project-related costs are reimbursed by the owner, and a fee is paid to cover profit and contractor home office overhead. This is also known as a time and materials contract. A guaranteed-maximum-price contract is a cost-plus contract in which the contractor agrees not to exceed a specified cost.

Project owners choose which method they want their contractors to price a project for a variety of reasons, including the completeness of the design, the complexity of the project, and the owner's experience in managing construction projects. Each of these pricing models also has a significantly different effect on jobsite cost accounting. The GMP pricing method is utilized for the hotel case study and examples in this book as it introduces the reader to a variety of cost accounting operations that the other pricing methods do not include, especially lump sum. There are many good project management and estimating books available, including *MOCP*, which include a more detailed discussion of these contract pricing methods, some of which is elaborated on in the next chapter on estimating.

Construction contracts

The construction contract is the most important construction document. It has significant impacts on how costs are accounted for, especially in an open-book contract. It is a legal document that describes the rights and responsibilities of the parties, for example, the owner and the general contractor. Five things must be aligned for a contract to exist:

- An offer to perform a service,
- An acceptance of the offer,
- Some conveyance, i.e. transfer of a completed building for money,
- A legal agreement, and
- Only authorized parties can sign the agreement, for example the officer-in-charge or chief executive officer (OIC or CEO).

The intent of the contractual agreement and the contract documents describe the completed project and the terms and conditions the parties (usually the project owner and the general contractor) must adhere to in order to accomplish the work. There typically is not a detailed description of the sequence of work or the means and methods to be used by the contractor in completing the project. The contractor is expected to have the professional expertise required to understand the contract documents and select appropriate subcontractors and qualified craftsmen, materials, and equipment to complete the project safely and achieve the specified quality requirements. For example, the contract documents will specify the dimension and sizes of structural steel columns and beams, but will not provide the design for erection aids, such as erection bolts or erection seats or hoisting and safety considerations. The contract documents usually include at least the following five essential elements. They are listed here in what had been considered a relative order of precedence, but today most see them as complementary; all of these documents must work together:

1. The contract agreement,
2. Special or supplemental conditions,
3. General conditions,
4. Technical specifications, and
5. Drawings.

The terms and conditions of the relationship between the primary parties are defined solely within the contract documents. These documents should be read and completely understood by the contractor before deciding to pursue a project. They also form the basis for creating a project estimate and schedule. To manage a project successfully, the project manager must understand the organization of the contract documents and the contractual requirements for his or her project. This knowledge is essential if the PM has any expectation of satisfying the expectations of his or her company executives as well as that of the project owner.

The contract agreement itself is either a generic standard template or a specially prepared document to suit a specific project owner or a specific project. Most government agencies use standard formats for developing construction contract documents. Federal and state agencies typically have standardized general conditions and contract language. Many private owners use contract formats developed by the American Institute of Architects (AIA) or the new family of construction contracts called the ConsensusDocs headed up by the Associated General Contractors of America (AGC) and other contractor associations. Contracts should not be signed until they have been subjected to a thorough legal review. This is to ensure that the documents are legally enforceable in the event of a disagreement and that there is a clear, legal description of each party's responsibilities. Standard contract agreements have typically been developed by attorneys and have been tested in the legal system and deemed fair to the parties that willfully execute them. The contract documents have a significant impact on the responsibilities of the jobsite project team and effect accounting processes. Specific requirements are contained in the general conditions and special conditions of the contract. A particular project could utilize many potential combinations of delivery, procurement, pricing methods, contract formats, and contractor organizational structures as shown in Figure 3.5.

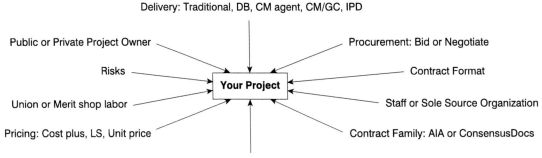

Figure 3.5 Delivery options

There are several different accounting implications or rules the contractor must adhere to which will be defined in the contract documents. Federal projects which utilize prevailing wages require additional documentation as discussed in Chapter 18, 'Taxes and audits.' Lump sum contracts are often briefer than negotiated contracts. Since they are closed-book they do not discuss reimbursable versus non-reimbursable costs. Open-book projects discuss at length what is cost-reimbursable and non-reimbursable. Reimbursable costs may include the project manager and cost accountant or cost engineer if they are stationed at the jobsite. Reimbursable and non-reimbursable costs are defined in Articles 7 and 8 respectively in AIA contract A102 for cost-plus projects with a guaranteed maximum price. Labor burden rates and the owner's right to audit the contractor's books are defined in the open-book project contract along with allowable equipment rental rates and maintenance costs. Closed-book lump sum projects are not auditable and are not subject to these clauses and articles. Both closed- and open-book contracts will define change order markups in their contract as well. Excerpts from AIA A102 Articles 7 and 8 pertinent to how the project will be estimated and accounted for follow.

Article 7, Costs to be reimbursed:

7.2.1: Wages of construction workers employed at the site (i.e. carpenters and laborers)

7.2.2: Salaries of supervisory personnel stated at the site (i.e. superintendent, project manager, and jobsite cost accountant)

7.2.4: Taxes, insurance, benefits for personnel (i.e. labor burden)

7.5.2: Rental charges for equipment

Article 8, Costs not to be reimbursed:

8.1.1: Salaries of personnel stationed at the office (i.e. CEO and accountants and marketing staff)

8.1.2: Expenses of the principal office, other than the site office

8.1.3: Home office overhead and general expenses

Project management

Project management is both a process and an individual or a company which includes the application of knowledge, skills, tools, and techniques to the many activities required to complete a project successfully. In construction, project success generally is defined in terms of safety, quality, cost, schedule, and document control. The construction team's challenge is to balance quality, cost, and schedule within the context of a safe project environment and document all aspects accordingly. Safety is of utmost importance to companies and individuals in the built environment industry, and sacrificing any aspect of safety to improve quality, schedule, or cost performance is unethical, dangerous, and unacceptable to all participants.

This section will examine project management and cost accounting from the perspective of the general construction contractor. Other project managers typically are involved in a project representing the owner and the designer and most of the principles presented here apply to them as well. The context of this book will be that of PMs and cost engineers for mid-sized commercial general contractors. The principles and techniques discussed, however, are equally applicable on residential, industrial, and infrastructure or heavy-civil construction projects, as well as specialty subcontractors.

The project manager and superintendent share leadership oversight of the contractor's jobsite project team and are responsible for identifying project requirements and leading the team in ensuring that all are accomplished safely and within the desired budget and time frame. The focus of the superintendent is on the field installation side and the PM on the office and field management side. To accomplish this challenging task, the PM must organize his or her project team, establish a project management and cost control system that monitors project execution, and resolve issues that arise during the course of construction. In a complete book or course on construction management, there are many other non-financial CM tools that a PM will utilize in managing a project, including requests for information (RFIs), submittals, meeting notes, and others. Many of those tools are depicted in the Construction Document Cloud, Figure 3.6.

There are five major phases of a construction project which overlap with the accounting cycle introduced in Chapter 1 and the cost control cycle which will be introduced in Chapter 8. These phases include planning or preconstruction, start-up, control, close-out, and post-project analysis. During project *planning*, the project manager, superintendent, and upper management evaluate specific risks that are associated with the project, particularly those related to safety, cost, quality,

Figure 3.6 Construction Document Cloud

and schedule. Risk analysis and risk management are critical skills essential to successful project management. The PM develops the organizational structure needed to manage the project and the communications strategy to be used within the project management organization and with other project stakeholders. Material procurement and subcontracting strategies are also developed during the planning phase, and along with jobsite material management, will be discussed in Chapter 11.

During *start-up* the project manager and superintendent mobilize the project management team, educate them regarding the project and associated risks, and conduct team-building activities. The project management office is established, and project cost accounting systems are initiated. Vendor accounts are established, and materials and subcontract procurement initiated. Project cost, schedule, safety, and quality control systems are established to manage all aspects of project execution.

Project control is a broad encompassing term which involves 'controlling' or 'managing' the project during construction, interfacing with external members of the project team, anticipating risks by taking measures to mitigate potential impacts, and adjusting the project schedule to accommodate changed conditions. The project manager monitors the document management system, quality management, cost control, and schedule control systems, making adjustments where appropriate in conjunction with his or her project team which includes the superintendent and the jobsite cost accountant. He or she prepares or reviews many performance reporting tools to look for variances from expected performance and takes action to minimize impacts to overall project success.

Project close-out includes not only completion of the physical construction of the project, but also submission of all required documentation to the owner, and financial close-out. The project manager must pay close attention to detail and motivate the project team to close out the project expeditiously to minimize jobsite overhead costs. Management of jobsite overhead is discussed in detail in Chapter 5 and close-out, particularly financial close-out, is discussed in Chapter 16.

Post-project analysis, if conducted, involves reviewing all aspects of the project to determine lessons that can be applied to future projects. Issues such as: estimated cost versus actual cost, planned schedule versus actual schedule, quality control, supplier and subcontractor performance, construction equipment choices, effectiveness of communications systems, and work force productivity should be analyzed. Many contractors skip this phase, and simply go to the next project. Those who conduct post-project analyses learn from their experiences and continually improve their procedures and techniques. The lessons-learned analysis may be included as part of the PM's close-out or with an internal company audit which will be introduced in Chapter 18.

General contractor organizations

The size and structure of a project's jobsite organization depends upon the culture and style of the corporation, the size of the project and its complexity and contract terms, and the jobsite location with respect to other projects or the contractor's home office. The cost of the project management organization is considered jobsite overhead and must be kept economical to ensure the cost of the contractor's construction operation is competitive with other contractors. The jobsite overhead costs are also referred to as indirect costs. The goal in developing a project management organization is to create the minimum organization needed to manage the project effectively. If the project is unusually complex, it may require more technical people than would be required for a simpler project. If the project is located near other projects or the contractor's home office, technical personnel can be shared among projects or backup support can be provided from the home office. If the project is located far from the contractor's home office, such as with our case study, the jobsite office must be self-sufficient.

General contractors organize their project management teams in either one of two models. In one type of project management concept, estimating, accounting, and scheduling are performed in the contractor's home office by staff specialists as illustrated in Figure 3.7. In an alternative organizational structure, estimating, accounting, and scheduling are the project manager's responsibilities, as illustrated in Figure 3.8. This sole-source style of organization is the one used

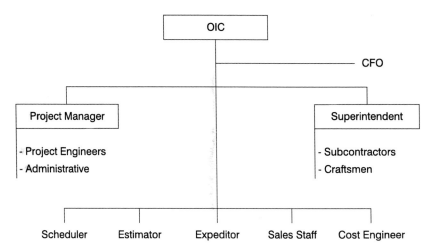

Figure 3.7 Staff organization chart

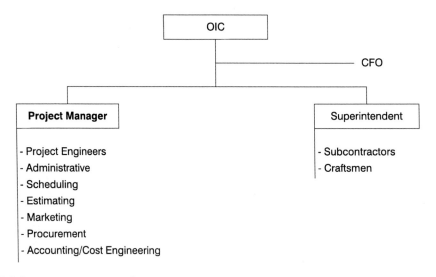

Figure 3.8 Sole-source organization chart

throughout this book. Both the PM and the superintendent will report to an individual or individuals within the contractor's upper-management or officer corps. The PM and the superintendent need to work together as a cohesive team, each with his or her areas of specialization, in order for the project to be successful. The choice of project management organizational structure is made by the contractor's corporate officers such as the CEO.

Once the project management organization is selected for a specific project, the project manager identifies the individuals to be assigned to each position. Most people will come from within the contractor's organization, but some may be hired externally. Selection of project team members from inside the construction firm may be made by the PM or may be made by senior company managers. Once all the team members have been selected, the PM must forge them into a cohesive team. Team building, both with internal and external partners, is a responsibility of the general contractor's PM and superintendent and is a focus of a more advanced independent study on construction leadership.

Contractor team member responsibilities

Construction team member responsibilities will vary from company to company and from project to project. The *officer-in-charge* (OIC) is the principal official within the construction company who is responsible for construction operations. He or she generally signs the construction contract and is the individual to whom the client turns in the event of any problems with the project manager. The OIC may also be the vice president for operations, chief operations officer (COO), district manager, senior PM, or may be the construction company owner or chief executive officer. In the case of a small contractor, these may all be the same person who may also be the PM and superintendent as discussed in some of our later examples.

The *chief financial officer* (CFO) is often an equity partner within the construction firm. This individual is responsible for all of the accounting operations of the company including taxes, accounts payable, accounts receivable, and internal financial statements including the balance sheet and income statement. The general contractor's CFO often communicates direct with the Internal Revenue Service, the client's CFO, and the CFOs of the subcontractors as well. The CFO may also be involved with running the contractor's independent equipment rental operations or real estate development arms if applicable. These concepts are introduced later in the book. The CFO may have other titles including bookkeeper, accountant, finance manager, and others depending upon the size of the company.

The *project manager* reports to the officer-in-charge and has overall responsibility for completing the project in conformance with all contract requirements; this is typical for the general contractor, project owner, designers, and subcontractor PMs as well. He or she organizes and manages the contractor's project team. The focus of this book will be on the functions of the GC's PM in conjunction with the jobsite cost accountant. Specific responsibilities of the PM include:

*   Coordinating and participating in the development of the project budget and schedule;
*   Developing a strategy for executing the project in terms of what work to subcontract;
*   Negotiating and finalizing contract change orders with the owner and subcontractors;
*   Submitting monthly progress payment requests to the owner;
*   Managing financial close-out activities, and others.

The *superintendent* is responsible for the direct daily supervision of construction field activities on the project, whether the work is performed by the contractor's direct craftsmen or those employed by subcontractors. On larger projects this is delegated to and accomplished by last planners, or those directly responsible for the work, such as assistant superintendents and/or foremen. Specific general contractor superintendent responsibilities include:

- Planning, scheduling, and coordinating the daily activities of all craftspeople working on the site;
- Determining the building methods and work strategies for construction operations performed by the contractor's own work force (means and methods);
- Ensuring all work performed conforms to contract requirements;
- Ensuring all construction activities are conducted safely, and others.

Project engineers or *field engineers* (PEs or FEs) typically report to the project manager and are responsible for coordinating daily details relating to field construction and documentation. On small projects, the PE's responsibilities may be performed by the PM. On large projects there may be multiple PEs. Specific PE responsibilities include: processing submittals and requests for information and maintaining associated tracking logs; preparing contract documents and correspondence and maintaining the contract file; and reviewing subcontractor invoices and requests for payment.

If the project is of sufficient size or remotely located or of a contract nature which warrants a *jobsite cost accountant* or *cost engineer*, that person will have similar responsibilities and background as the project engineer, and may be called a project engineer, but with primarily a cost focus. The cost accountant is located on the jobsite for the hotel case study and exercises utilized throughout this book. The jobsite cost accountant will work closely with the project manager to record costs, prepare monthly pay requests, and forecast future expenditures. The jobsite cost accountant will also work closely with the superintendent and help prepare cost control work packages, amongst other financial responsibilities.

Foremen are last planners who report directly to the superintendent and are responsible for the daily direct supervision of craftsmen on the project. The construction firm will assign foremen for work that is performed by the company's own construction craftsmen. Foremen for all subcontracted work will be assigned by each subcontractor. Specific responsibilities include: coordinating the layout and execution of individual trade work on the project site; verifying that all required tools, equipment, and materials are available before work commences; and preparing daily or weekly time sheets for their crews.

Risk analysis

Construction is a risky business for a variety of reasons and is proven out by the high number of construction firm failures each year. To minimize the potential for financial difficulty, a contractor should analyze each potential project to determine the risks involved and whether or not the potential rewards justify acceptance of the risk exposure. Risk management is part of any financial study and involves risk identification, measurement, and mitigation strategies.

The sources of *external risk* on a project may involve things such as unusually adverse weather, material cost inflation, owner's inability to finance the project, limited availability of skilled craftspeople or subcontractors, bankruptcy of subcontractors, incomplete design documents, project location, theft and vandalism, safety, and project complexity. The contractor needs to forecast the likelihood of such risks, the range of possibilities, and the impact of each on the contractor's ability to complete the project profitably. Projects with increased risk require a larger fee in return for the risk accepted than those with less risk.

Internal risks will also be identified and must be managed both at the corporate and project levels. The three most common internal risks are unrealistic cost estimates, unrealistic construction schedules, and ineffective project management. Project management short-comings include cost and schedule control, material management, and subcontractor coordination. Risk from internal theft is minimal and can be managed as described throughout this book.

Contractors must adopt strategies to minimize the potential of these problems occurring. Often the basic issue to be addressed is the selection of qualified people to manage the project, particularly the project manager and superintendent. The output of a risk analysis process results in a decision about whether or not to pursue a project, the amount of fee and contingency to include in the bid or cost proposal, whether or not to joint-venture with another firm, the portions of work to subcontract, and the type and amount of insurance to purchase. Resolutions of these issues are also discussed throughout this book.

Summary

There are four major project delivery methods; the primary differences among them are the relationships between the three primary project participants. Owners select contractors by one of two methods, bidding or negotiating. Public owners often are required to openly bid projects and private owners may choose whatever procurement system they are comfortable with. In both cases, contractors may go through a prequalification process which shortens the list of firms bidding or proposing. Project owners also determine the method by which the project is to be priced by the general contractor. The primary pricing methods include lump sum, unit prices for heavy-civil projects, and cost plus fee projects. Many cost-plus projects also have a guaranteed maximum price which financially protects the project owner on the high side but provides cost savings opportunities if the contractor under-runs its estimate.

The construction contract describes the responsibilities of the owner and the contractor and the terms and conditions of their relationship. A thorough understanding of all contractual requirements is essential if a project manager expects to complete the project successfully. Contracts are either standard or specifically prepared documents. Standard contracts generally are preferred because they have been tested in and out of the legal system.

The contractor's project manager is the leader of the jobsite construction management team. He or she is responsible for managing all the activities required to complete the job on time, within budget, and in conformance with quality requirements specified in the contract. The major phases of a construction project are: planning, start-up, control, close-out, and post-project analysis. Contractors establish project management organizations to manage construction activities. The project team typically consists of a PM, superintendent, project engineer, foremen for self-performed work, and administrative support personnel depending upon project size and complexity. Larger or remote projects, especially those with open-book accounting requirements, may also have a cost accountant or cost engineer located at the jobsite.

Construction is a risky business, and project managers must carefully assess the risks associated with each prospective project. Once the risks have been identified, risk management strategies must be developed. In some cases, the risks are too great, and the project should not be pursued. In other cases, the risk can be mitigated by obtaining a joint-venture partner or hiring subcontractors or with increased fees.

This chapter was not meant to be a comprehensive coverage of construction management or project management, but just a brief introduction to a variety of terms and processes that connect to a study of construction cost accounting and financial management. We hope we have sparked an interest and the reader is suggested to look to a more comprehensive resource such as *Management of Construction Projects, A Contractor's Perspective (MCOP)*, by Schaufelberger and Holm, for a more in-depth discussion.

Review questions

1. What are the advantages for an owner to utilize a traditional delivery method and a lump sum price? What are the disadvantages for a contractor for the same scenario?
2. What are the disadvantages for an owner to utilize a DB delivery method and a cost plus percentage fee pricing scenario? What are the advantages for a contractor for the same scenario?
3. Why would a GC choose to locate their accounting operations at a) the home office, and/or conversely b) the jobsite office?
4. Why would a GC choose to bid a project? There are many potential reasons.
5. When is it acceptable for a client to let a project to bid? There are many potential reasons.
6. Why would a client choose one of the following pricing methods?
   a. Lump sum,
   b. Unit-price,
   c. Cost-plus, or
   d. GMP.
7. What are the differences between an agency CM and an at-risk CM?
8. What criteria would a project owner use to choose a GC on a bid project, and/or conversely a negotiated project?
9. How much more responsibility does a GC PM have in a sole-source organization versus a company which relies heavily on its staff organization?
10. Why do lump sum contracts, such as AIA A101, not differentiate between reimbursable versus non-reimbursable costs as does the AIA A102?

Exercises

1. Why would a project owner choose to utilize one of the DB extended delivery methods, such as DBOM?
2. Draw an organization chart for a) Evergreen Construction Company and the Olympic Hotel and Resort case study project, or b) a project you have been working on outside of the classroom. Include at least ten companies, positions, and/or individuals. Make whatever assumptions are necessary.

3. In addition to the engineers and consultants listed as BE participants in this chapter, list one additional firm or type of firm.
4. Subcontractor types were not discussed specifically in this chapter nor are a focus of this book. List three which would be on a) a typical commercial project, b) a single-family house, and/or c) a heavy-civil project.
5. Add one additional responsibility for each of the GC team members described in this chapter.
6. Match up the five project phases with the cost control cycle in Chapter 8. What occurs with what?
7. Prepare a table matching the four primary delivery methods with the best choices of:
   a. Procurement: Bid or Negotiate,
   b. Five design phases, and/or
   c. Four pricing methods.

4

Introduction to estimating

Introduction

Cost is one of the most critical project attributes that must be controlled by the project manager (PM), superintendent, and cost engineering team. Project costs are estimated to develop a construction budget within which the project team must build the project. Cost estimating is the process of collecting, analyzing, and summarizing data in order to prepare an educated projection of the anticipated cost of a project. All project costs are estimated in preparing bids for lump sum or unit-price contracts and negotiating the guaranteed maximum price (GMP) on cost-plus contracts. This project budget is the first step and becomes the basis for the cost control system that will be discussed in Chapters 8 through 11. The goal of this chapter is not to reproduce all of the information that is available in estimating-specific textbooks, but to highlight some of the major issues in developing cost estimates. Before a study of construction cost accounting processes can begin, a baseline estimate must be established, from which the project team can 'account' to.

Project cost estimates may be prepared either by the project manager and his or her project engineers or by the estimating department of the construction firm. When possible, the PM, cost engineer, and superintendent should be responsible for developing the estimate or at a minimum, work as integral members of the estimating team. Their individual inputs regarding constructability and their personal commitments to the final estimate are essential to assure not only the success of the estimate, but also the ultimate success of the project. One of the first assignments for the estimating team is to develop a responsibility list and to schedule the estimate. The estimating process should be scheduled for each project beginning with the dates for the pre-bid or pre-proposal conference and finishing when the bid or proposal is due. With these milestones established, a short bar-chart schedule should be developed which shows each step and assigns due dates to the estimating tasks depicted in the Estimating Process Figure 4.1.

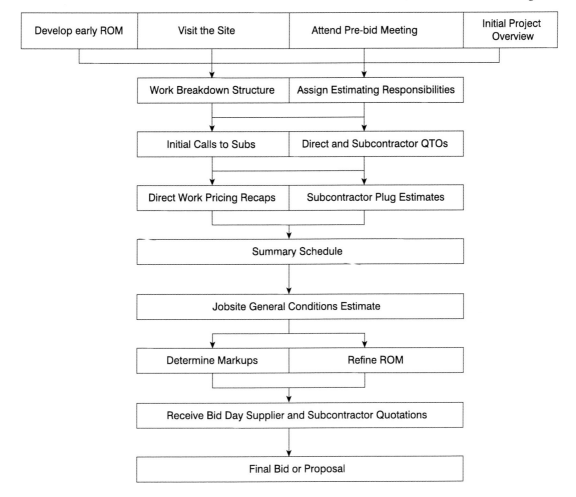

Figure 4.1 Estimating Process

Types of cost estimates

Many outside of the construction industry view all contractors' estimates the same, in that they are all 'firm bids' and all contractor-produced estimates are completely detailed and accurate. This of course is a misconception and largely dependent upon document completion. There are several different types of cost estimates. Conceptual cost estimates are developed using incomplete project documentation, while detailed cost estimates are prepared using complete drawings and specifications. Semi-detailed cost estimates are used for guaranteed maximum price contracts and have elements of both conceptual and detailed estimates as discussed further later. The estimating strategy or approach is different with each of the three main types of estimates and the level of detail will differ as well. The accuracy of an estimate is directly proportional to the accuracy of the documents and the time spent on preparing an estimate also follows those same lines. All estimates have major elements or cost categories, some of which require a more detailed effort by the estimator than others. The revenue equation, and subsets of this equation, is shown here and will be referred to throughout this study on construction cost accounting.

$$\text{Revenue} = \text{Cost} + \text{Profit}$$
$$\text{Construction cost} = \text{Direct cost} + \text{Indirect cost}$$
$$\text{Direct cost} = \text{Labor} + \text{Material} + \text{Equipment} + \text{Subcontractors}$$
$$\text{Indirect cost} = \text{Jobsite general conditions} + \text{Home office general conditions}$$

Early estimates may be developed by contractors or architects on limited information and produced quite quickly. These estimates are not expected to be 'firm' nor are they necessarily accurate. They should be referred to as *budgets, schematic estimates*, or *conceptual estimates*. A conceptual set of drawings can be estimated quite quickly using square foot (SF) of floor unit prices, assembly prices, subcontractor plugs or budgets, and percentage add-ons for general conditions to produce a rough-order-of-magnitude (ROM) budget. Completion of the schematic design (SD) phase is also usually followed by a contractor-generated, or estimating consultant-generated, budget estimate, but this is not a firm bid. Most items within a budget are by definition allowances or plugs. Budgets are the least accurate estimate type and should carry substantial contingencies, such as 10–20%. Budgets are also developed only by seasoned estimators, whereas the junior estimator or cost engineer will assist with quantity take-offs (QTOs) associated with detailed lump sum bid estimates.

A contractor will also develop an early and quick budget from even a detailed set of drawings as soon as they come through the door. This estimate is referred to as a ROM estimate and is utilized to determine if the project is the right size for the contractor to pursue and which of its estimating or management staff would be best suited to work on the subsequent detailed estimate. Another early estimate prepared by the project owner, which is often a real estate developer, is a pro forma, which is the topic of the last chapter in this book.

A *detailed estimate* takes the longest time to prepare, costs the contractor the most in personnel resources to complete, and produces the most accurate final figure. Usually drawings and specifications which are 90–95% complete are associated with detailed estimates. This would be consistent with completion of the construction document (CD) design phase. Detailed estimates are required for projects which are lump sum bid, such as for a public waste water treatment plant.

Although lump sum bids are customarily associated with public works projects, clients on privately financed projects may also solicit lump sum bids during a slow economy. In a busy economy though, project owners may only interest contractors with negotiated requests for proposals. Clients soliciting lump sum bids are primarily interested in just the bottom-line price and assume all contractors can deliver the project with comparable levels of quality, schedule, and safety. As shown in Figure 4.1, the detailed estimate process includes quantity take-offs, pricing recaps, early subcontractor plug estimates, jobsite general conditions, competitive subcontractor and supplier quotations, and markup choices.

Guaranteed maximum price estimates and resultant guaranteed maximum price (GMP) contracts are a hybrid of budgets and detailed estimates. GMPs are often prepared on negotiated projects after completion of the design development (DD) phase. Detailed estimates are produced for scopes which have been adequately designed and specified, such as structural concrete and steel, utilizing measured quantities and unit prices. Subcontractor plugs or allowances are included for areas not yet designed, but competitive subcontractor bids are factored wherever possible. For work which is not fully designed, the contractor will use assembly costs such as $/SF or allowances. Contingencies are more prevalent in GMP estimates than detailed estimates and may amount to 2–5%. Projects with a GMP are usually performed open-book and any resultant savings are shared between the client and general contractor (GC), such as 80% to the client and 20% to the GC. A definition of costs which are reimbursable and those which are non-reimbursable must be established. Also reimbursable costs, which are considered jobsite general conditions, versus those general conditions which are considered part of the home office and therefore included with fee are significant and should be established in the contract before work commences. Evergreen Construction Company negotiated a GMP estimate and entered into an AIA A102 contract with Northwest Resorts for the example case study referred to throughout this book.

The process to prepare a detailed estimate is the subject of the balance of this chapter, which includes the work breakdown structure, quantity take-offs, pricing direct and subcontracted work, jobsite general conditions, and culminating with the estimate summary and associated markups.

Work breakdown structure

The process of developing an estimate was illustrated in Figure 4.1. The first step of gathering information is the foundation for the process. As the estimator proceeds through the process, information continues to be analyzed and summarized and refined, until eventually there is only one price left, the final estimate or bid. The work breakdown structure (WBS) for the project is an early outline of the significant work items that will have associated cost or schedule considerations. This includes scopes of work such as concrete walls, exterior paving, ceramic tile, electrical, etc. Before detailed take-offs and pricing are prepared, the estimator should develop this general picture of all of the work that will be included on the WBS. One of the first steps in the estimating process is to perform a brief document overview. The estimator should slowly leaf through the drawings and specifications and develop a good understanding of the type of project and the building systems that are included. The estimator should not start quantifying or pricing any work items until this overview is complete.

Figure 4.2 is an abbreviated work breakdown structure example for the case study project. An expanded WBS is included on the eResource as a solution to an exercise. This list is a good reference to use throughout the estimating process as well as a final checklist to review again prior to finalizing the estimate. This early WBS is not to be considered a final product, just a good first step. Cost codes are a theme throughout this study of accounting and financial management and the WBS is a good place for the estimator to start using them – even if only with two-digit Construction Specifications Institute (CSI) divisions or six-digit specification sections. The WBS will continue to evolve throughout the estimating, scheduling, and subcontract buyout processes. As the estimator dives deeper into the project, there will be several more detailed subsequent levels of the WBS.

Construction tasks which are to be self-performed will require a more intense quantity take-off and pricing effort. It is appropriate at this early work breakdown structure phase for the estimator to separate direct versus subcontracted scopes of work. Some of the types of work commercial

Figure 4.2 Work
breakdown structure

```
                    Evergreen Construction Company
                      Work Breakdown Structure

                       Olympic Hotel and Resort

        CSI    Description
        ─────────────────────────────────────────────
         2     Demolition
         3     Concrete
         4     Masonry
         5     Structural Steel
         6     Carpentry
         7     Waterproofing
         8     Doors and Windows:
                    Windows:
                              Storefront Windows
                              Punch Windows
                              Door Relites
                              Un-framed Mirrors
                    Doors, Frames, and Hardware:
                        Door Frames:
                              HM Door Frames
                              Grout HM Frames
                              Wood Door Frames
                              Aluminum Door Frames
                        Door Leafs:
                              HM Door Leafs
                              Solid Wood Leafs
                              Hollow Wood Leafs
                              Pocket Doors
                              Plam Door Leafs
                              Bi-fold Doors
                        Door Hardware Sets
                        Access Doors
         9     Finishes
        10     Specialties
      Continued....
```

general contractors may perform with their own direct crews include: concrete formwork, structural steel erection, rough carpentry, finish carpentry, doors, specialties, and accessories. These are scopes of work often installed direct by the GC's carpenters, laborers, and ironworkers. Decisions regarding which scopes of work to self-perform and which to subcontract are based on several criteria:

- Subcontractors will be used if the specialized tradesmen needed are not employed by the GC.
- The reason subcontractors are also known as specialty contractors is that they specialize in a specific scope of work, such as windows, and should be expected to know that work better than would other contractors.
- If there are problems with the quality of installation, a subcontractor is required to repair the work without increase in cost to the GC.

- The subcontractor may have estimated a fixed price which is less than what the GC has estimated therefore the subcontractor would be absorbing the pricing risk.
- Subcontractors may have craftsmen and equipment available and the GC does not.
- There are risk mitigation arguments on both sides as to which, the subcontractor or the GC, can perform the work faster, safer, cheaper, and with better quality than the other.
- In the case of the Olympic Hotel project, the jobsite is three hours from Evergreen's normal place of business. It would be difficult and expensive to transport their direct crews or house them during the week. Therefore Evergreen may subcontract more work on this project than they might if the project was near their office; within the greater Seattle area.

Another tool the construction team might utilize at this stage is a project item list, which is essentially a work breakdown structure with columns indicting direct labor and material from that of subcontractors. A live Excel version of the project item list is included on the eResource.

After the estimator, project manager, and superintendent have decided which scopes of work to subcontract, subcontractors should be notified about the project. Today this is likely done electronically with an invitation to bid and an access code to drawing files. It is a good idea to have the subcontractors working to assist with the estimate in parallel with the general contractor developing its own estimates for self-performed work. When subcontractors are notified they will ask the estimator questions regarding specifications, quantities, and materials. The cost engineer or estimator should be informed at that time about specific subcontractor scopes but should be cautious about providing too much detail to the subcontractor. General contractors do not want to place themselves in the position that subcontractors have based their prices solely on information the GC estimators have given them. Each subcontractor should develop a completely independent estimate.

The work breakdown structure, if properly developed including sufficient detail and cost codes, can also function as a preliminary 'bill-of-materials' from which the superintendent and cost engineer can issue short-form purchase orders and place material orders.

Quantity take-off

After the work breakdown structure is developed, the estimator takes material quantities directly off the drawings. The estimator prepares quantity take-offs by conducting detailed measurements and counts of each item of work that the team selected to be self-performed. The take-off process is a time-consuming step in the estimating process as shown in Figure 4.1, and is a crucial step in preparing any detailed construction estimate.

The quantity take-off process starts with the work items that will be constructed first, the foundations. This is typical for commercial or civil or residential projects. This order will accomplish several tasks. First, it will force the estimator to think like the builder – the foundations are built before the floor systems. This is also why it is important to have the superintendent involved early in the estimate creation. Organization of the estimate in this fashion will later assist with the schedule development and will aid with the development of the project cost control systems. All the work items should be taken off prior to pricing, but some estimators perform both QTO and pricing simultaneously – another advantage offered by computer estimating tools. Material quantities are recorded on QTO sheets similar to the one for the Olympic Hotel as shown in Figure 4.3.

Evergreen Construction Company
Quantity Take-off Sheet

Project:	Olympic Hotel, Job #422					Date:		03/25/2019
Owner/Loc:	NW Resorts, LLC					Estimator:		Chris Anderson
						Est #:		1
Div #:	05, Structural Steel							

Ref	Description	Qty (ea)	L (ft)	W (ft)	H (ft)	Total LF	#/LF	Total #
S2.1	Columns: 6 x 6 x 1/2" HSS	40			12.5	500	35.24	17,620
S2.4	Girders: W24 x 30	24	30			720	30	21,600
S2.4	Lintels: 6 x 6 x 1/2"<	48	30			1,440	19.6	28,224
	Total Members:	112	Ea					
	Total Weight							67,444
	Add 10% for Gussets and Plates							74,188
	= Tons @ 2000 #/Ton							37.1
								Tons
Notes:	Miscellaneous steel and metal decking are quantified on separate QTO sheets,							
	Anchor bolt QTO with foundations, Steel safety cable QTO with rough framing							

Figure 4.3 Quantity take-off

Counted and measured quantities of materials must be mathematically converted to purchasable units that match up with standard unit pricing. For example, concrete is measured in length multiplied times width multiplied times height or thickness or depth which produces cubic feet, but it must be divided by 27 to produce cubic yards. Likewise structural steel is purchased by the ton, wood framing by the thousand board feet, and plywood floor and wall sheeting by individual sheets. The amount of waste or lap to apply will vary with the material, installer, project, and estimator. Allowing between 5% and 10% is common and included on the quantity take-off. Purchasing enough, but not too much, material is important to maintain labor productivity. Items such as nails, glue, caulking, and rebar tie-wire can be estimated or allowed for but determining exact quantities is difficult and outside of Pareto's 80-20 rule, where 80% of the cost is included in 20% of the activities. The amount of time and cost the estimator expends on these small items exceeds the value of the materials. Allowances are usually sufficient at this stage.

Quantity take-off sheets for subcontracted work are prepared similarly, but by system or assembly and need not be as detailed as that of the general contractor's direct work. For example, the GC will take-off sheetrock walls, including metal studs and taping, by the square foot or linear foot of wall system, and not independently quantify sheets of drywall, metal studs, and taping mud. Order-of-magnitude (OM) estimates will be developed for subcontracted work to be replaced later when competitive quotes are provided by the subcontractors.

Pricing

There is no one exact estimate for any project but some estimates are more correct than others. Adjustments in pricing, subcontractor and direct labor strategies, jobsite overhead structures, and fee calculations are individual contractor decisions that will determine the most correct estimate for those conditions at that time.

Pricing recapitulation sheets (recaps) are developed for each system, or CSI division, that is utilized. The estimator should not start pricing until all of the materials have been taken off and recorded on the quantity sheets. After all of the quantity take-off has been completed, the estimator should begin pricing labor and direct materials. All of the recapped material quantities are circled or highlighted and brought forward from the QTOs and entered onto the pricing recap sheets. An example of a pricing recap sheet is shown in Figure 4.4.

Pricing self-performed work

The men and women who work out on construction jobsites with their tools and their hands should not generically be referred to as 'construction workers.' Rather they have a specific trade or craft, such as a carpenter and plumber, and are referred to as tradesmen or craftsmen. They have been trained and specialized in one craft such as electricians installing conduit and wire. They are also not all to be referred to as 'construction laborers' as the laborer is a specialized craft who places concrete along with other responsibilities. There is not yet an accepted gender-neutral term for tradesmen, craftsmen, journeymen, or foremen; 'workers' is too generic, so this book will stay with male gender terms for simplicity. Although tradesmen and craftsmen may be abbreviated as 'trades' or 'crafts.'

Direct labor productivity is the most difficult item to estimate and is therefore considered the most risky item in the estimate. Contractors can only guestimate how many hours it will take to perform a task; there are no rules and they can only manage, not control, actual field productivity as will be discussed in Chapter 8. Often a general contractor will review the amount of direct labor in an estimate and use this figure as some basis for determining overall project risk and therefore the appropriate fee to apply. Labor productivity is estimated as man-hours (MH) per unit of work, such as 10 MHs per door. This is referred to as unit man-hours or UMH. This system allows for fluctuation of labor rates, union versus open shop choices, geographic variations, and time. If it takes 8 MHs per ton to install foundation reinforcement steel in Atlanta, it probably takes the same in Aberdeen. Appropriate wage rates can be applied for the specific project. The best source of labor productivity is from the estimator's in-house database. Each estimator or contractor should develop their own database from previous estimates and as-built labor history. Other sources of labor productivity rates include published databases or reference guides, such as RS Means' *Building Construction Cost Data*.

It is important to round calculated man-hours off on the pricing recap sheets. Fractional extended MHs should not be retained. Partial MHs are difficult to schedule and monitor against for cost control, let alone explain to the superintendent why he or she has 2.8 MHs to install hotel unit kitchen appliances and not three hours. Estimating is not an exact science; it is a combination of math and good judgment. The MHs are totaled at the bottom of each pricing page. The total system hours will be valuable information later for scheduling and cost control.

Evergreen Construction Company
Pricing Recap Sheet

Date: 03/25/2019
Estimator: Chris A.
Estimate #: 1

Project: Olympic Hotel and Resort, Job #422
Location: Rainforest, WA
Architect: Gateway Design
Division: 05, Structural Steel

Code	Description	Qty	Unit	Material Rate	Material Cost	Labor UMH	Man Hours	Wage Rate	Labor Cost	Equip Rate	Equip Cost	Total Cost
055134	HSS Columns	40	EA			1.5	60	41	2,460			$2,460
050524	Anchor Bolts	w/concrete foundation estimate										$0
051200	Girders	24	EA			2	48	41	1,968			$1,968
051515	Lintels	1,440	LF			0.115	166	41	6,790			$6,790
052567	Cable Railing	w/wood framing estimate										$0
055900	Shake-out	37	Tons			0.33	12	41	501			$501
050575	Plumb & Align	37	Tons			0.5	19	41	759			$759
050526	Bolt-up	37	Tons			0.5	19	41	759			$759
014545	Safety	37	Tons	100	3,700	0.33	12	41	501	90	3,330	$7,531
015550	Hoisting: Crane	3	wks			40	120	38	4,560	1,750	5,250	$9,810
015446	Connections: 40' Boomlift	3	days							350	1,050	$1,050
015670	Welder	3	wks							150	450	$450
015677	Compressor	3	days							118	354	$354
051250	Buy Steel	37	Tons	2,200	81,400							$81,400
Notes:	Miscellaneous metals and metal decking are priced on separate pricing recap sheets				$85,100		395		$15,836		$10,434	$111,370

Wage Check:
Labor $ / Hours = 40.09 $/HR.   Checks? **Yes**

System Check:
Total $ / TN = 3010 $/TN   Checks? **Yes**

Figure 4.4 Pricing recapitulation

The wage rates that get applied are determined by the company, location, union agreements (if applicable), and the type of work to be performed. Public construction project wage rate values may also be governed by Davis-Bacon or prevailing wage rates which are established by the Department of Labor. An example wage schedule for Evergreen Construction is shown later in Table 18.2. On non-prevailing wage projects only bare (paid to the employee) wage rates are included on the pricing recapitulation sheets. Labor burdens or labor taxes are included on the estimate summary page. This line item will add between 30–60%, depending upon craft, on top of the labor estimate and covers costs such as workers' compensation, union benefits, unemployment, FICA (social security), and medical insurance. These percentages fluctuate with time and location. The estimator should check with his or her company's chief financial officer and current accounting data before applying a labor burden markup percentage. Labor burden, including labor taxes and labor benefits, are discussed in detail in Chapter 18.

Some estimators will use crew wage rates which combine journeymen with foremen with apprentices. This often averages out at the journeyman's wage and is a common and acceptable practice. Some will use blended wage rates which factor carpenters and laborers and cement masons and ironworkers all together. Others will use loaded wage rates which combine bare wages (that which the craftsman receives) with labor burden. A detailed well-organized estimator and cost engineer will keep different crafts and direct wages separate from burden to facilitate better cost control efforts and more accurate as-built estimate development. This detailed approach also facilities easier open-book accounting audits.

The best sources of material unit prices are from suppliers, not from databases. The process to procure construction equipment rental rates is similar to material pricing. Suppliers are invited to provide prices at the same time subcontractors are – early in the process. Each material item that will be purchased by the general contractor should receive a current market-driven material or equipment quote. Price quotations should be received in writing on the supplier's letterhead. A live example of a telephone bid proposal form is available on the eResource. If telephone quotations are received, company-consistent forms should be used for recording material pricing, although pricing on the vendor's letterhead is always preferred. Many material quotes are received electronically. Other sources of material prices include in-house historical costs or previous estimates. Historical prices are not as reliable as current quotes, but it is better than leaving an item blank. The third choice for material prices are databases or reference guides. These sources suffice for unusual items for which local quotes cannot be obtained or as allowances or subcontractor plugs or order-of-magnitude estimates in early budget development until competitive prices are received.

Pricing subcontracted work

The best source of subcontractor pricing is from subcontractors. They are the ones who will ultimately be required to sign a contract and guarantee performance of an established scope of work for a fixed amount of compensation. Subcontractors are invited to bid on a project as one of the early steps in the estimating process. Just as general contractors market to project owners, they should also treat past, current, and future subcontractors with respect so that they receive complete and competitive future bids. Some subcontractors may also provide GCs with early budgets for their scopes as place-holders until bid day quotes are refined. Additional ways to treat subcontractors fairly are to assure timely monthly payments and release of final retention after close-out as will be discussed in Chapters 14 and 16 respectively.

General contractors should estimate subcontracted scopes in-house to refine day-one rough-order-of-magnitude budgets and to check the reasonableness of subcontractor bids once received. These in-house GC estimates for subcontracted work are referred to as order-of-magnitude estimates or plug estimates. If one low bid of $432,500 for roofing comes in and another higher bid of $900,000 is received, the GC can feel comfortable throwing the high bid out because their pre-bid plug estimate was $450,000. The reverse may be true on another system, if the GC's estimate was nearer the high bid. Once a reliable subcontractor price is received, the GC's plug estimate should be replaced with the subcontractor's price. The estimator should not assume that the firms which specialize in a particular scope, such as landscaping, are all in error and that his or her early in-house plug figure is the most correct one to be used on bid day. A bid tab spreadsheet is customarily developed to aid in the analysis of comparing subcontractor and supplier bid day pricing.

Jobsite general conditions

There are many different uses of the term general conditions which will be expanded on in upcoming chapters. Jobsite general conditions costs are project-specific, not general or generic. Although estimating direct labor costs is one of the most difficult and riskiest tasks of the construction estimating team, estimating jobsite general conditions is also difficult and risky. Site-specific general conditions costs are also referred to as jobsite administration costs.

Because so many line items in the jobsite general conditions estimate template are time-dependent, the contractor must have some idea of a proposed project's construction duration beyond that provided by the project owner or architect in the request for quotation. Schedules are often required to be submitted with bids or cost proposals, but even if not required, an in-house schedule is necessary to support preparation of the jobsite general conditions estimate. Utilizing past experience, coupled with the estimated direct work man-hours and with the superintendent's input, it is a fairly simple process to prepare a 20–40 activity summary schedule. The project team will rough out a schedule based upon these hours, the project's complexity, and early subcontractor input on durations and deliveries. This schedule is used to develop overall durations for estimating site administration and equipment rental for the jobsite general conditions estimate. It is not necessary to determine if it will take the project team exactly 320 working days to build the building. But it is important to know that this structure will take approximately 16 months to construct (not 12 or 24) given the site conditions, project complexity, long lead material deliveries, anticipated weather, and available manpower. A summary schedule for the Olympic Hotel and Resort case study project and an abbreviated general conditions estimate are included with Chapter 5. A detailed construction schedule and a detailed general conditions estimate for the case study project are also included on the eResource.

Estimate summary

Many additional elements are necessary to complete the construction cost estimate. An estimate summary page will be utilized to gather all of the direct work estimates, subcontractor pricing, general conditions, and markups. An estimate summary form, similar to other estimating forms, should be developed and used uniformly throughout the firm. It may require slight modification for any one project but its consistency is important to provide the project estimating team with a quick and comfortable overview prior to and during bid day. With all estimating, construction management, and cost accounting forms, consistency is important throughout any construction firm. Consistency helps reduce errors and therefore improves productivity and profits.

In the case of a bid job, the estimate summary should be filled out as much as possible the day prior to bid day. A pre bid-day estimate summary should be prepared as a refinement of the first rough-order-of-magnitude estimate that was developed the day the bid or proposal documents came into the contractor's office. As the estimator works to complete the estimate, he or she should use the most current and relevant pricing information. All of the pricing recap pages are brought forward by posting the values on the summary sheet. The estimator should have someone else check the math as it is here that many gross errors may occur. The computer is an excellent and standard tool to use at this time and throughout the estimating process. Printing hard copies may seem archaic to some in the technology age, but often this is only where blatant errors may literally 'jump off the page' at the estimator and/or his or her superiors.

On the bottom of the estimate summary, below the subtotal of the direct, indirect, and subcontracted costs, are a series of percentage markups including the fee. The fee is comprised of home office overhead and profit, or 'OH&P'. Home office overhead costs include items such as accounting, marketing, and officer salaries. These costs usually are a relatively fixed figure based upon staffing for one fiscal year. An average-sized general contractor may need to generate a minimum fee of 1% to 4% (that is 1% to 4% of their total annual volume) to cover these costs. In addition to covering the indirect costs, the contractor desires to earn a profit on each project. The desired profit plus the home office overhead indirect cost is the fee. Fees can range anywhere from 2.5% to 7.5% for commercial work and often 10% and higher for civil and residential projects due to the increased direct labor risk. Smaller commercial contractors also require higher fees on their projects due to smaller annual volumes. If the GC self-performs much of the work, they also have a proportional increased risk and require a higher fee. On cost-plus contracts, conceptual cost estimates may be used to establish the guaranteed maximum price, but home office overhead costs still need to be estimated to develop the fee proposal. There are many percentage add-on markups which will show up on any estimating summary page. The type of contractor (subcontractor, supplier, commercial GC, civil, residential) also affects which markups and percentages are appropriate. Follows are a few of the more common ones:

- Labor burden is not really a markup but is a cost of the work, especially direct work. Contractors which use loaded wages will include burden above the subtotal line with the work and others will factor it below the line with the balance of the other markups.
- General liability insurance is volume related. The insurance rate will vary significantly between contractors depending upon their size and safety records. This markup generally ranges from less than 1% for large commercial and civil contractors to 2% for smaller GCs and subcontractors.

Evergreen Construction Company
1449 Columbia Avenue
Seattle, WA 98202
206-447-4222

GMP Summary Estimate
Olympic Hotel and Resort, Project #422

CSI Div	Description	Direct L. Hours	Labor	Material	Equip.	Subs	Total
01	Jobsite General Conditions (1,2)	5,272	$804,280	$516,566	$170,850	$171,087	$1,662,783
02	Demolition (3)	0	$0	$0	$0	$0	$0
03	Concrete	27,848	$1,141,775	$2,209,857	$199,980	$1,275,000	$4,826,612
04	Masonry	0	$0	$0	$0	$150,000	$150,000
05	Structural & Misc. Steel	2,000	$82,000	$128,103	$142,000	$59,000	$411,103
06	Rough and Finish Carpentry	7,694	$300,060	$234,567	$78,889	$2,505,677	$3,119,193
07	Thermal and Moisture	221	$7,505	$8,200	$7,200	$1,371,909	$1,394,814
08	Doors and Windows	432	16848	452897	3775	811420	$1,284,940
09	Finishes	0	$0	$0	$0	$2,337,066	$2,337,066
10	Specialties	325	$12,675	$90,551	$1,350	$52,509	$157,085
11	Equipment	390	$15,210	$192,000	$0	$51,340	$258,550
12	Furnishings	0	$0	$0	$0	$96,687	$96,687
13	Special Construction	0	$0	$0	$0	$0	$0
14	Elevators	0	$0	$0	$0	$435,000	$435,000
21	Fire Protection	0	$0	$0	$0	$442,700	$442,700
22	Plumbing	0	$0	$0	$0	$1,211,379	$1,211,379
23	HVAC and Controls	0	$0	$0	$0	$720,629	$720,629
26	Electrical	0	$0	$0	$0	$1,286,702	$1,286,702
27	Low-Voltage Electrical	0	$0	$0	$0	$332,500	$332,500
31	Sitework	103	$3,500	$7,279	$3,434	$1,414,098	$1,428,311

	Subtotals:	44,285	$2,383,853	$3,840,020	$607,478	$14,724,703	$21,556,054
	Labor Burden on Direct L	50.00%	$1,795,653			$897,827	$22,453,881
	Labor Burden on Indirect L	30.00%	$588,200			$176,460	$22,630,341
	Subcontractor Bonds	w/Sub $				$0	$22,630,341
	State Excise, B&O Tax	0.75%				$169,728	$22,800,068
	Liability Insurance	0.95%				$216,601	$23,016,669
	Builder's Risk Insurance	by Owner				$0	$23,016,669
	Fee	5.00%				$1,150,833	$24,167,502
	Contingency	1.50%				$362,513	$24,530,015
	GC Bond	Excluded				$0	$24,530,015

Total GMP: $24,530,015

Notes: 1) See separate detailed estimate
2) GC hours are for direct labor only
3) Completed by a previous contractor

Figure 4.5 Summary estimate

- Taxes, including material sales tax, business tax, or excise tax should be job-costed but are dependent upon the city, county, and state.
- If the project is competitively bid with relatively complete documents, the amount of contingency applied is usually zero. Most GCs account for the contingency, if any, with their choice of fee on competitive projects. Stated contingencies will show up on negotiated projects which have incomplete documents.
- Performance and payment bond rates are also annual volume and company performance related but the project bond costs are project-specific. They generally do not appear on negotiated projects. On larger lump sum projects, they may be required, especially for public and civil work, and the cost will range dependent upon the size of the project and the past performance of the contractor. The bond cost is customarily 'below-the-line,' that is below the base bid, or is an alternate add-on and not included in the quote in a lump sum bid.

The focus of upcoming Chapter 6 is on home office overhead and profit, which equals the fee. Some of these other markups will be discussed there as well. The final bid or proposal value is determined on bid day. This is the final step in Estimating Process Figure 4.1. Subcontractor quotations will be received, plugs will be removed, and the pre bid-day budget totals revised and replaced with hard bid quotes. The final bid generally is approved by the officer-in-charge of the construction firm. The total figure must be submitted on the form specified in the instructions to bidders to the client before the specified bid time. A summary estimate for the Olympic Hotel case study is shown in Figure 4.5. A detailed ten-page estimate is included on the eResource.

Summary

The three major types of estimates prepared by contractors include budget or conceptual estimates, detailed estimates associated with competitive bids, and semi-detailed estimates presented as guaranteed maximum prices on negotiated projects. One additional type of estimate prepared by the project manager and cost engineer near project completion is the as-built estimate. This estimate records actual craft hours and material costs with actual quantities and feeds back into the contractor's estimating database. As-built estimates will be discussed in future chapters as well.

Similar to constructing a building, estimating is a logical process consisting of a series of steps. The first being project overview to determine if the project is going to be pursued. Once decided a work breakdown structure should be outlined which will assist with both the estimate and construction schedule development. The quantity take-off step is a compilation of counting items and measuring volumes. Pricing is divided between direct labor pricing, material pricing, and subcontract pricing. Labor cost is computed using productivity rates and current local labor wage rates. Material and subcontract prices are developed most accurately using competitively bid supplier and subcontractor quotations. Jobsite general conditions cost is a job cost and is schedule dependent.

Home office overhead is combined with desired profit to produce the fee percentage. The fee calculation on any specific project varies dependent upon several conditions including company volume, market conditions, labor risk, and resource allocations. The first major lesson to be learned in estimating is to be organized. If proper organization and procedures are utilized, good estimates will result. The second is to estimate and estimate a lot. Practice and good organization will eventually develop thorough and reasonable estimates. The third lesson is to maintain and use historical databases of cost and productivity results from previous projects.

This chapter was not meant to be a self-contained source for estimating, but rather a kick-off to some of the cost accounting and financial management activities performed by the jobsite team. They could not properly do their jobs without a detailed construction estimate. Live Excel versions of the estimate forms used in this chapter as figures, and others, are available on the eResource. There are many complete and detailed textbooks on estimating, including *Construction Cost Estimating, Process and Practices*, by L. Holm et al. (2005). The interested reader should look to resources such as that for a more thorough coverage of this important financial management building block.

Review questions

1. What is the difference between a GMP cost estimate and a detailed cost estimate?
2. What are some of the risks a contractor faces when developing a cost estimate? There are several.
3. Why would you, as a PM or superintendent or cost engineer, want to be involved in the estimate (or schedule) development for your project?
4. Why should a contractor prepare its own preliminary construction schedule during the estimating process?
5. How does the 80-20 rule relate to estimate development?
6. Where do the most accurate estimate unit prices come from?
7. How do the terms of a GMP contract affect a contractor's approach to estimate development?
8. What level of design (SD, DD, CD) is most commonly associated with a) lump sum estimates, b) GMP estimates, and c) budget estimates?
9. If a GC's pre-bid plug for cabinets is $100,000 and they do not receive a sub bid on bid day, what should they do?
10. How can a contractor minimize estimate errors?
11. Which of these estimate types (budget, GMP, and lump sum bid) are the most accurate and which includes the highest contingency rates?

Exercises

1. Using Figure 4.1 and the case study firms and individuals you can garner from several chapters throughout this text, and assuming a bid date three weeks from today, develop an estimating assignment list for each team member, including due dates. Include a 'check person' for each major deliverable.
2. Expand the WBS in Figure 4.2 for the Olympic Hotel and Resort case study to the second and third levels for two CSI divisions such as 03 and 09.
3. Without looking ahead, which of the major cost elements of the estimate do you believe deserves the greatest cost control focus and why that one?

5

Jobsite general conditions

Introduction

There are several different terms used to describe general conditions (GCs) in construction as well as different meanings for that term, including:

- American Institute of Architects (AIA) A201 general conditions (contract document),
- Construction Specifications Institute (CSI) specification divisions 00 and 01,
- Overhead costs,
- Indirect costs,
- General requirements,
- General and administrative costs,
- Distributable costs,
- Administrative costs, and
- Operational costs.

They do not all mean exactly the same thing to all of the built environment participants, so it is important for the construction project manager (PM) to be clear about the application for the term they are using. This chapter focuses on *jobsite general conditions*. The next chapter will focus on home office overhead and profit (OH&P). For many, it is easiest to include the prefix *jobsite* or *home office* to differentiate these very important aspects of construction management and their relation to estimating, accounting, and financial management.

As discussed in the last chapter on estimating, the major elements of a construction estimate, which are a subset of the revenue formula, include:

Construction cost = Direct cost + *Indirect cost (Jobsite general conditions)*
Direct cost = Direct labor + Material + Equipment + Subcontractor costs

Direct costs reflect actual field construction, such as forming concrete footings, erecting structural steel, hanging and taping gypsum wall board, connecting light fixtures, and hundreds if not thousands of other activities necessary to complete a construction project. The results of these efforts, be they labor or material, are visible during the course of construction and/or when the project is completed. But the results of home office or jobsite general conditions are *not directly* visible or measurable in the project. These efforts are therefore aptly described as *indirect costs* and they are *distributed* over the course of all of the work.

This chapter will introduce many of the different elements of jobsite general conditions, the process for estimating each of those elements, and development of a complete jobsite general conditions estimate. There are strategies utilized by the contractor to develop that estimate as well as manage it. It is the contractor's project manager and superintendent who are directly responsible for managing these costs. GCs at the jobsite level are very time-dependent; if the project lasts longer they cost more. Change orders and delays impact those costs but the impacts are sometimes difficult to quantify and are not always reimbursable.

Elements of jobsite general conditions

There are many different ways to determine the scope of jobsite general conditions for an estimate. It can be as simple as taking a previous estimate and making slight modifications for the new project. Although the term *general* is used when describing this work, every project is specific and the GCs estimate should also be project-specific. Sophisticated contractors have lengthy templates that they use to pick and choose items which apply to a specific project. One of the first things the estimator will do is to scour the client's contract and the front end of the specifications (CSI divisions 00 and 01) for management requirements that he or she must incorporate into this project. Different forms of contracts, as was introduced in Chapter 3, may have a significant impact on estimating and accounting. Jobsite conditions include laydown, security, and a hoisting plan and all have impacts on jobsite GCs. Then by picking and choosing potential estimate items from a lengthy template, the estimator can begin the process of developing a customized cost estimate. There are typically four major categories in the GCs template. Eventually the estimate may be collapsed into an abbreviated single page or two, again job dependent, and these four categories may end up all grouped together. What follows is a list of the four categories and a few sample items which will be found within each. A complete four-page blank Excel GCs template is included on the eResource.

1. Administrative expense:
   a. Project manager's wages,
   b. Superintendent's wages,
   c. Project engineer's wages,
   d. Surveying (see Strategies section later),
   e. Jobsite accountant's wages (see Strategies section later), and others.
2. Equipment:
   a. Superintendent's pickup truck,
   b. Compressors and welders if not charged to the work,
   c. Forklift,

    d.   Tower crane (if applicable),

    e.   Personnel and material hoist (if applicable), and others.

3. Temporary construction:

    a.   Jobsite office trailer,

    b.   Installation and maintenance of temporary site utilities,

    c.   Parking accommodations for craftsmen,

    d.   Temporary fencing, and others.

4. Operational costs:

    a.   Percentage add-ons and markups (if not on the summary estimate page),

    b.   Construction utility bills,

    c.   Drinking water and chemical toilets,

    d.   Flaggers (important for heavy-civil work),

    e.   Temporary cleanup,

    f.   Final cleanup (if not with subcontracts), and others.

Some of these items may be included with the direct work portion of the estimate as described in the Strategies section later. Some of the items may be excluded from this specific project or may be provided by others such as the project owner. It is important not to eliminate any of the items from the template as then one would not know if they had been intentionally excluded or simply missed. Specifically excluded items should be left on the template and appropriately annotated. Rarely, if ever, would a construction loan and corresponding interest charges be included in the contractor's estimate. There is not a line item for loan interest in our template, but one could be added if the estimator felt it was required. Cash flow is the topic of Chapter 13, pay requests in Chapter 14, and the developer's pro forma in Chapter 19, which are all topics related to a construction loan.

Development of jobsite general conditions estimates

After the project manager and chief estimator have customized the general conditions estimate template for their specific project, they can begin populating it with quantities and units and pricing. As shown in the Estimating Process in Figure 4.1, the GCs estimate is not started until most of the other detailed estimating work has been accomplished. Many of the items in the GCs estimate are time-dependent, so a construction schedule is necessary and will be developed as follows:

- Total all of the direct construction work hours,
- Superintendent will review the hours, and with estimated crew sizes, determine durations,
- Subcontractor input on durations is very helpful,
- A construction schedule should be developed, even if not required by the client's request for quotation or request for proposal (RFQ or RFP), and
- The contractor's total duration should be compared to that of the client's if provided in the contract documents.

It is not necessary to develop a fully detailed construction schedule at this time unless the contractor is the only one submitting a negotiated proposal and is very confident of award. After

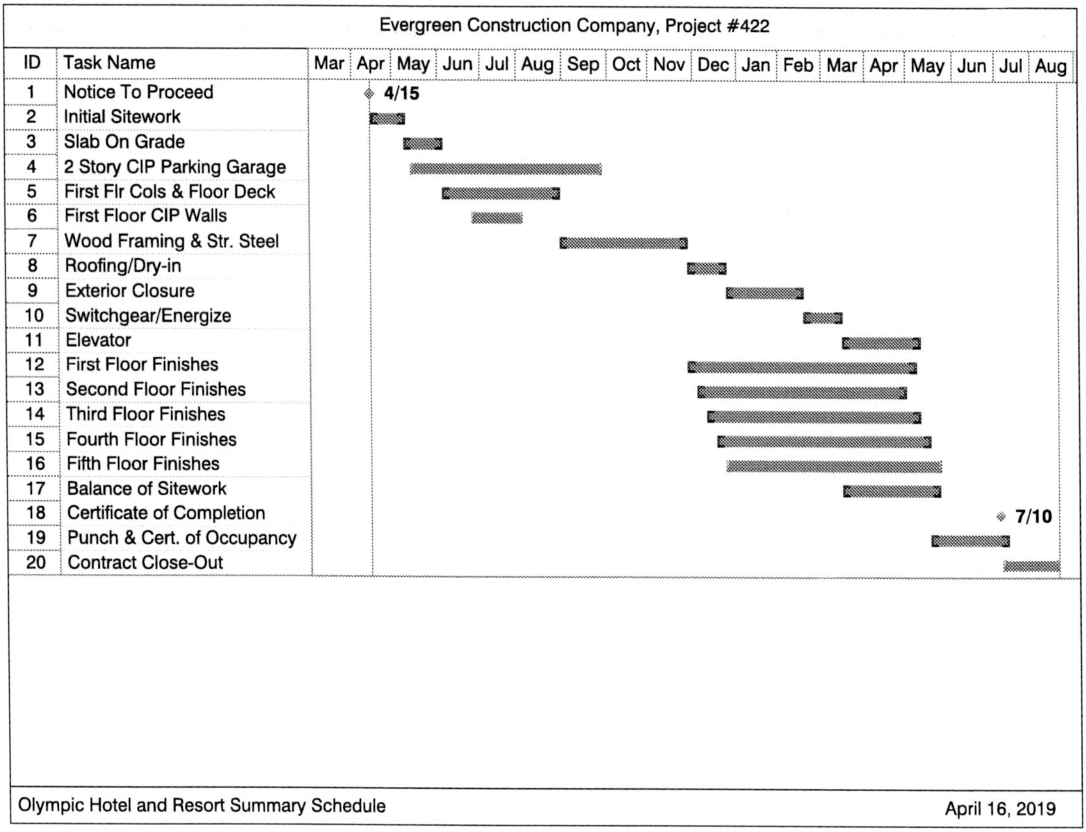

Figure 5.1 Summary schedule

award in a competitive bid or negotiated scenario, then a detailed schedule will be necessary which will be in excess of 100 line items long if not closer to 1,000. A summary schedule is often requested with competitive bids and proposals and will be sufficient to develop the general conditions estimate. A summary schedule for the hotel case study is included here as Figure 5.1. A fully detailed schedule is included on the eResource.

Now that the estimator knows this is a 16-month project, many of the general conditions cost items can be calculated. For example, as it relates to our case study, the superintendent will be full time on the project and he earns a monthly salary of $10,000 per month. The project manager and project engineer (PE) and administrative assistant will also be full time. The jobsite accountant will split her time between standard PE activities and cost engineering activities. The superintendent's pickup truck and forklift and two jobsite trailers are all estimated for the full project duration as well as many other line items. An abbreviated GCs estimate for the case study has been included here as Figure 5.2. An expanded version is also available on the eResource.

Other items in the general conditions estimate may be dependent on the cost of some elements of the work or on an anticipated cost of the entire project. An example is small tools which are not provided by the craftsmen but estimated at 5% of the direct labor estimate. This is another reason why the direct cost of the work was necessary to have been calculated before the GCs cost was developed. Other items, if applicable for this specific project and not included on the estimate summary, are calculated on a percentage of the total cost of the work. Estimators do not typically prepare quantity take-offs for GCs as they will for direct work.

Pricing should not be obtained from published estimating databases such as RS Means for general conditions. GCs need to be job-specific as stated earlier, but also are company-specific. One firm may pay their accountant $1,200 per week and another $1,500 per week. One firm may supply their superintendents with pickup trucks to take home at the end of the day and others may expect superintendents to provide their own trucks in turn for increased compensation included in their salaries.

		Evergreen Construction Company **Jobsite General Conditions Estimate** (Abbreviated)					
Project: Olympic Hotel and Resort Owner: NW Resorts, LLC					Estimator: Estimate # Date:		Chris Anderson 1 03/25/2019
			Direct Labor		**Material/Equip/Subs**		Total
Description	QTY	Units	U. Price	Cost	U. Price	Cost	Cost
Project Manager	68	wks	2,500	170,000		0	170,000
Project Superintendent	68	wks	2,350	159,800		0	159,800
Project Engineer (1.5 PEs)	102	wks	1,200	122,400		0	122,400
Jobsite Cost Accountant (1/2 as PE)	34	wks	1,200	40,800		0	40,800
HO Safety Inspector (1 x weekly)	68	wks	300	20,400	100	6,800	27,200
HO QC Inspector (1 visit weekly)	68	wks	300	20,400	100	6,800	27,200
Scaffold Subcontractor	1	LS				105,748	105,748
Pickup Truck (2 EA)	32	mos			1,050	33,600	33,600
Compressor	w/cost of the work					0	0
Welder	w/cost of the work					0	0
Forklift	16	mos	6,680	106,880	3,950	63,200	170,080
Small Tools	1,795,653	DL $			5.6%	100,000	100,000
Job Office (2 EA)	32	un/mo			1,000	32,000	32,000
Dry Shacks/Tool Vans (3 EA)	16	un/mo			1,800	28,800	28,800
Temporary Lighting	125,000	SFF			0.15	18,750	18,750
Radios, 4 ea	64	un/mos			75	4,800	4,800
75kw Generator for Crew Shack	16	mos			2,750	44,000	44,000
32kw Generator for tools, 2 each	32	un/mos			1,800	57,600	57,600
Fences	1,620	LF			4	6,480	6,480
Street use Permit	Excluded	mos			0	0	0
Flaggers	624	Hrs	33	20,280	0	0	20,280
Periodic Cleanup	68	wks	1,300	88,400	0	0	88,400
Continued...							
GRAND TOTAL JOBSITE GENERAL CONDITIONS:				$804,280		$858,503	$1,662,783
LABOR BURDEN ON SALARIED LABOR:		Carried on estimate summary page					$0
BURDEN ON CRAFT LABOR:		Carried on estimate summary page					$0
TOTAL JOBSITE GENERAL CONDITIONS:							**$1,662,783**
% OF TOTAL BID @	$24,530,015		CHECKS?	**YES**			6.8%
MONTHLY RATE @	16 months		CHECKS?	**High, but out of town and fast, OK**			$103,924

Figure 5.2 Jobsite general conditions estimate

An experienced estimator will develop a quick general conditions estimate utilizing the processes described here, then print it out and let it sit for an hour or so. Then they take a red pen to it and mark it up and input the revisions and go through the editing process several additional times. Drafts of the estimate should be reviewed with the superintendent to get his or her buy-in. Some contractors allow their home office chief estimator to prepare the project-specific GCs estimate, but the PM and superintendent should play a major role, if not prepare it themselves. At a minimum they should review and approve the scope and time and pricing.

There is no exact final figure the general conditions will total, but they generally range from 5% on larger projects to 10% on smaller projects. Residential projects and heavy-civil projects are on the higher end. The amount of direct work a contractor has, which requires more intense supervision and management, also affects the final cost. Negotiated open-book projects will have a higher GCs cost than closed-book bid projects. The total percentage should be noted on the bottom of the GCs estimate page as well as alongside the GCs line item on the total estimate summary page as was shown in Figure 4.5. The GCs estimate for Evergreen Construction Company (ECC) on the Olympic Hotel and Resort case study project totaled 6.8% of total contract value. This may be a little on the high end, considering this is a $24.5 million project, but because of its out-of-town location, this is a very acceptable amount.

Many estimators also relate general conditions cost to an average dollars per month calculation. Different types of projects have significant impacts on the monthly cost, for example if a tower crane or a personnel hoist is required. This monthly average should also be noted on the GCs estimate. If an estimate is significantly higher or lower in either percentage or monthly cost compared to other past projects, the estimate should be re-visited and either modified or the estimator should understand the differences because of the unique aspects of this specific project. The estimator should independently audit his or her own work before finalizing and posting the GCs total to the estimate summary. An easy way to do this is to simply scan down the sheet and look at the largest and smallest items; do they appear correct?

Labor burden (labor taxes plus labor benefits) was introduced prior and will again be addressed with taxes and audits. There are different schools of thought about whether burden should be combined with the cost of direct wages and included above the subtotal line with the work, or below the line as a percentage markup. In the case of a general conditions estimate, craft labor for scopes such as cleanup or equipment operation has been combined with indirect labor such as a safety officer and jobsite accountant. Salaried labor receives a much smaller burden markup than does craft labor, for example 30% compared to 50%, so they need to be calculated separately. Craft labor also varies depending upon union preferences. An experienced client may also ask for stipulated wage rates and labor burden markups on the bid form or in the proposal. Some estimators will include the labor burden for GCs on the GCs estimate page and others below the line on the estimate summary page. Regardless of the location, it is imperative that markup rates for craft and salaried labor burden be figured separately and that the burden is included in one location in the estimate, and not omitted, and not figured twice.

This discussion has been based upon the view that the contractor will prepare a detailed general conditions estimate and not just allow a percentage add-on, such as 8%, on top of the cost of the direct work including subcontractors. This would be the approach an estimator would take only with schematic budget estimates, but not detailed lump sum or guaranteed maximum price estimates prepared by experienced contractors. Smaller, less-sophisticated contractors may estimate GCs as a percentage add-on to the total estimate, but this is not reliable and not advised and cannot be monitored, controlled, and accounted for as is discussed later.

Strategies

Contractors can take several different approaches to the way they estimate and manage both their home office and jobsite general conditions. It is a misconception to think of GCs as *overhead* especially at the jobsite level. Business analysts will discuss reductions in overhead to lower costs and improve profits. Although this may be true for contractor home office GCs, it does not necessarily apply to the jobsite. These administrative and operational expenses are necessary to support the field construction. Spending additional jobsite GCs can actually save direct field costs and improve fee, which improves profit. An example would be spending additional costs on weekly jobsite cleanup which improves site access, material handling, labor productivity, and site safety conditions.

Design firms quote *fees* to their clients, which combines all of their overhead costs plus profit plus the cost of labor and material and equipment to prepare their design. When clients and designers evaluate a general contractor's jobsite GCs estimate, they think in terms of fee, but these expenses are not a fee. Some built environment participants believe contractors have hidden fees buried within their GCs estimates, but actually contractors tend to lose money on their GCs estimates, i.e. they spend more on management than what they estimated to support the field efforts, often due to schedule delays. Some construction project owners utilize a hybrid procurement system of requesting a lump sum fee (including home office overhead and profit) and jobsite GCs, along with customary RFP items such as resumes and company history. This is often the case with CM-at-risk or CM/GC delivery methods. Because of the focus contractors receive on their GCs estimates the percentage GCs amount to on top of the cost of direct work therefore becomes competitive. Contractors subsequently strategize on how GCs estimates are prepared and managed and develop ways to lower the estimate and potentially cost. Following are some of these strategies:

1. First, contractors will move home office GCs to jobsite costs wherever possible. Because fee equals home office overhead (HOOH) plus profits (Fee = OH&P), and the company's total net profit is really their focus, any reduction in HOOH and shifting cost to the projects, all the while keeping the same fee, improves profit. Article 7 of open-book AIA contract A102 defines what costs are reimbursable and part of the job cost and Article 8 defines non-reimbursable costs. A contractor may propose language changes or inserts to help move costs from the home office to the jobsite and have the client pay for them. Some examples of costs which could be shifted from the next chapter's discussion on home office GCs to the jobsite would include:
   - Assigning a senior project manager to the project as a PM;
   - Charging the contractor-owned equipment repair and maintenance to the job;
   - Locating the PM at the jobsite instead of the home office, along with his or her computer and automobile expenses or rent; and
   - Utilizing a jobsite accountant to perform much of the cost engineering normally done out of the home office.
2. For example, assume a contractor has an estimated $100,000,000 total company annual construction volume with 5% fee, which includes 2% HOOH and 3% profit. This means

the company on average spends 2% of $100,000,000 or $2,000,000 on HOOH and realizes 3% of $100,000,000 or $3,000,000 of profit. Their total 5% fee is therefore $5,000,000. If 25% of those HOOH expenditures could be relocated to the projects, this would result in a reduction of $500,000 from HOOH which would then cost $1,500,000 (now 1.5%) and with the same fee of $5,000,000 their new profit margin is $3,500,000, or 3.5%. This represents a $500,000 profit gain without doing any additional work.

3. Commercial general contractors typically subcontract out 80% or more of the work; heavy-civil contractors subcontract less and residential contractors (especially spec builders) close to 100%. Direct labor is the riskiest portion of the estimate for any contractor to control costs. The more work subcontracted, the less is the risk and fewer jobsite administrative expenses will be necessary to manage the work.

4. General contractors will require their PMs to shift as much of their GCs activities to subcontractors as is possible. In order for this to happen, these transfers of responsibility and cost must be both feasible from a managerial perspective and contractually acceptable by the subcontractors. There are many examples of how this can happen, including:

   a. Building envelope subcontractors provide their own scaffolding;
   b. Roofing contractors (and other subcontractors working on the roof such as mechanical and electrical) provide their own hoisting;
   c. Subcontractors provide their own forklifts;
   d. Subcontractors provide their own dumpsters;
   e. Include temporary heat in the heating, ventilation, and air conditioning subcontractor's bid package;
   f. Include temporary electrical power and lighting in the electrical subcontractor's bid package;
   g. Subcontractors typically already provide their own office trailers and tool vans;
   h. Each subcontractor contributes one laborer per week to clean up the site, regardless of what needs to be cleaned up; and/or
   i. Earthwork and utility subcontractors provide their own flaggers.

5. Another means the PM and superintendent have to reduce jobsite GCs costs include cost coding activities and expenses to direct work activities in lieu of GCs. Foremen and general foremen and assistant superintendents charged with supervising craft labor activities can be cost-coded to the work such as concrete formwork. Equipment examples of this include charging cranes and welding machines to structural steel erection and compressors to carpentry and other activities. By including equipment with the cost of the work and not with GCs, the contractor will have a complete picture for what that system of work costs for future estimate development. Application of GCs to work activities is the subject of activity-based costing discussed later in this book.

6. If surveying is performed by the contractor's in-house surveyors then this cost is included in the GCs as it is distributed over several activities. Whereas if professional licensed surveyors are contracted as a subcontractor, then this cost can be grouped along with other subcontract costs. Final cleanup is also listed as an option on the GCs estimating template, but if a subcontractor is hired to do this work the estimator can also group those costs with other subcontractors such as painters and glaziers and not include the subcontractor cost in the GCs. Some general contractors will also subcontract out flagging, scaffolding, window cleaning, and site security services and not include them in the general conditions. Some of

these subcontractors were left in ECC's GCs estimate for transparency with this construction-experienced client.

7. There are also several line items on the GCs estimating template from the operations section that can be placed on the estimate summary page and not with GCs such as insurance, permits, bonds, taxes, and others. Those items are generally referred to as 'below-the-line' markups and not part of the cost of the direct work which is shown 'above-the-line' on the estimate summary.

8. Because the estimate is time-dependent, the best way for a contractor to save money on its jobsite GCs estimate is to plan and estimate and facilitate a faster way to build the project. Saving one month from the client's stated schedule in the request for quotation is one reason this author/estimator was the successful low-bidder on multiple occasions.

Accounting of jobsite general conditions

The project manager and the project superintendent are the two most 'accountable' parties on a construction project, especially from a general contractor's perspective. Their roles have already been discussed in preparing the jobsite general conditions estimate and strategies for managing those costs. Home office GCs are managed by the CFO and CEO, not the jobsite team, as will be discussed in the next chapter. During the course of construction the jobsite team must also manage the jobsite GCs expenses the same as they would for direct labor, material, and subcontract costs. Even though the PM and superintendent are responsible for the financial success of the project, they often rely on the assistance of project engineers, cost engineers, and jobsite accountants for cost control and support activities. The accounting cycle was introduced as Figure 1.1. The following are some of the steps that the jobsite team will go through when accounting for the GCs estimates:

1. The estimate must be corrected before it is submitted into the cost control system. From a jobsite GCs perspective this might involve changing the durations of time-dependent activities and personnel if it is discovered that the project will take a month longer or shorter to construct. Also if any of the rates have changed, for example a project engineer was estimated as full time at $1,200 per week but a more senior PE is being considered for the project at three days per week with a wage of $1,500 per week.

2. If scopes are subcontracted or purchase orders are let then buyout values and cost codes from those documents must be entered into the accounting system. Some of the elements of jobsite GCs are subcontracted as discussed earlier.

3. During the course of construction, time sheets must be filled out on a weekly basis and cost codes assigned, both for craft labor performing indirect activities, such as the forklift operator, and indirect salaried labor such as the field administrative assistant. Wages for most jobsite salaried personnel will be journaled into the job cost history report by the home office accounting department.

4. Indirect material invoices must be cost coded on a weekly basis. It is imperative that the cost codes be accurately input on the invoices by the jobsite accountant or cost engineer and exactly match those set up in the cost control system.

5. Equipment rental invoices must also be cost coded on a monthly basis. This includes everything from the pickup truck to the copy machine. A watchful eye must be kept on the maintenance

cost of owned and rented equipment; this will also be discussed later in this book. In addition rental durations must be optimized: it is less expensive to rent a piece of equipment for a whole week than four days, the same for a month versus three weeks out of the month. Mobilization and demobilization costs can be expensive for equipment continually rotated in and out of the job.

6. Once monthly the PM will prepare a cost forecast for upper-management with the assistance of his or her cost engineer or jobsite accountant, along with input from the superintendent. Cost and fee forecasting will be discussed in detail in Chapter 8. In the case of jobsite GCs, this is again very time-focused. If the project has been spending $90,000 per month on GCs and there are five months left to go, then likely there is approximately $450,000 worth of GCs yet to spend, unless a specific item such as a tower crane is finished and has been demobilized.

7. There is not the same estimating risk of missing productivity rates or quantities with GCs as there is with direct labor and material. The largest risk in missing the GCs estimate is time. If the project is estimated for 16 months, but lasts 20 months, there is likely a 25% cost over-run. The PM and superintendent must keep a close eye on schedule control, as well as cost and quality and safety controls.

Impact of change orders, claims, and delays on general conditions

As indicated, over-running the schedule will have the most significant effect on jobsite general conditions costs. For work that is clearly drawn and specified, the construction team bears all of the scheduling risk. But when changes in scope occur beyond the contractor's control, this may have an adverse effect on the schedule. Since the development of the GCs estimate is largely time-dependent, any delay in the time necessary to complete the job will directly cause an over-run in these costs. Accounting for change orders is the topic for upcoming Chapter 15. Most change orders are due to minor discrepancies and, individually, do not adversely affect the schedule. It is difficult for a contractor to prove how each individual change order item delays the schedule unless each scope item was originally shown on the contract schedule which will cause the schedule to be thousands of activities long and very difficult for the contractor to manage, report, status, and update. This author/expert witness observed a publicly bid high school construction schedule comprised of 10,000 line items and wrapping around all four walls of the jobsite trailer. That schedule was developed by the general contractor just to prepare for, and prove, schedule impacts and submit an end-of-project claim, which they did.

Some contracts have clauses that allow the contractor to recover specific general conditions items on a change by change basis, if those items can be substantiated. In this case it is advantageous for the contractor to have prepared a detailed construction schedule and a detailed GCs estimate and not an 8% add-on below-the-line estimate. But many contracts today have a *no claim for delay* clause which prohibits any of the construction team from recovering GCs cost if the project is delayed by the owner or designer, weather, or other means beyond the contractors' control. The contractor can receive extra time, or days on the schedule, but no additional general conditions costs or recovery for lost fee potential.

The change order proposal summary page looks very much like an estimate summary page. See Figure 15.1. Very seldom will a client allow a percentage add-on for jobsite general conditions; and home office GCs are assumed to be included in the fee markup. Subcontractors, however, are more successful at receiving additional overhead on their change orders because the value of their work is less and the amount of direct labor they have, and associated risk, is more.

Summary

There are different uses for the term *general conditions*. This chapter has focused on jobsite GCs – those activities and personnel and processes necessary to successfully manage the construction work for both the contractor's own forces and that of the subcontractors. There are many generic GCs estimate templates, but the estimator and project manager and project superintendent must customize a GCs estimate for each specific project. Many of the line items in the estimate are time-dependent. If the same construction equipment and personnel are necessary to manage a 16-month project as a 32-month project, the longer project will likely cost almost twice that of the shorter project. There is no exact percentage the GCs estimate will total compared to the entire project cost, but within the range of 5% to 10% for a commercial general contractor is fairly common.

Because some negotiated projects require the general contractor to compete based on a quoted fee plus a jobsite general conditions estimate, the contractor will strategize on ways to reduce its jobsite GCs estimate and costs. This includes everything from moving costs to subcontractors and direct work or completing the project faster. Management and accounting for GCs costs are the responsibility of the jobsite team. The processes utilized of cost coding and forecasting GCs is similar to that of direct labor and material expenditures. Also because the GCs are so time-dependent, change orders, especially the accumulation of multiple smaller changes, or one single large change order, can disrupt the schedule and adversely affect the chance of meeting the GCs budget. A detailed schedule and detailed GCs estimate is one of the best ways to recover change order specific GCs costs.

Review questions

1. List two other terms for 'jobsite GCs'.
2. Why is there not just one standard GCs estimate template that can be used for each project?
3. What are the largest individual labor and material and equipment GCs items in the hotel case study estimate? Are these typical for all projects?
4. Earlier it was proposed a crane should be charged to the direct work if possible such as to steel erection. Why is a tower crane then included as a standard item in the GCs estimate template included on the eResource?
5. What type of project would warrant a jobsite cost accountant?
6. The abbreviation GCs has been used a lot in this chapter for general conditions. What is that abbreviation also used for?

Exercises

1. Research AIA A102 and in what specific articles and paragraphs would the GC insert jobsite reimbursable costs?

2. Explain why it might be acceptable for a contractor to have a GCs estimate percentage out of the standard range, for example a) 3%, or b) 12%.

3. Other than trying to keep costs down, why would an owner not want a contractor to include a standard GCs percentage add-on for every change order, regardless of scope?

4. Should a contractor be required to provide a credit for GCs if scope is reduced but the schedule remains constant?

5. Looking at the long-form GCs template on the eResource, are there any additional items you as an estimator would have included on the hotel case study estimate?

6. Other than the examples listed with 'Strategies' earlier, how might you suggest a contractor save on jobsite GCs cost?

7. Assume the client for the hotel case study has added one month to the project before the contract was negotiated and is allowing a corresponding change to the jobsite GCs estimate and total contract value. How does this change:
   a.   The total GCs cost?
   b.   The total contract cost?
   c.   The percentage of GCs to total contract?
   d.   The GCs cost per month?

8. Perform the same calculations as in Exercise 7, but now assume the general contractor has figured a way to build the project one month faster without changing the original contract value. What would be the new projected fee?

9. Forklifts are customarily one of the most expensive GCs equipment expenses. Prepare an argument for a) why a dedicated forklift operator should be estimated and included in the GCs estimate and/or conversely b) why there should not be an operator in the estimate.

10. Looking ahead at the list of home office overhead items in Chapter 6, which ones might be moved to the jobsite for a) a small project which accounted for only 5% of the firm's volume and/or conversely b) a large project which accounted for 50% of the firm's annual volume?

11. Is the 'no claim for delay' clause fair for contractors? Why do sophisticated clients insert this? How might a compromise be achieved?

6

Overhead and profit

Introduction

In the last chapter, jobsite general conditions was discussed in detail. Management of jobsite general conditions is the responsibility of the project manager (PM) with assistance from the superintendent and the jobsite cost accountant. This chapter turns the focus back to the home office and introduces home office general conditions. Even though the project team has little to no input on management of home office issues, it is important that PMs and superintendents understand how home office overhead (HOOH) impacts the project. In this case, HOOH is added to the desired profit to produce fee as reflected in the following equations. Fee is a jobsite focus, as fee is what is left over after all of the construction costs are subtracted from the contract value or revenue. Fee is necessary from each construction project in order to pay for HOOH. Each project must cover its proportional share of HOOH.

> Revenue = Construction cost + Fee
> Fee = Home office overhead + Profit (OH&P)

This chapter discusses the creation of the home office general conditions (GCs) or overhead estimate and management of that estimate which is a chief financial and executive officers' (CFO and CEO) responsibility. It is very unlikely that most project managers will ever see their corporate GCs estimate, unless they become an officer in the company. The construction company must produce enough revenue from its jobsites to pay for the home office GCs. If the total fee from the jobsite exactly equals the home office GCs costs, then the company has just broken even. It is the ultimate goal of each project team to provide sufficient fees, beyond the breakeven point, to return a net profit which can be returned to equity partners or invested back into the company. Producing enough fee to break even is important, but it is not enough.

Creation of the general conditions estimate

When are general conditions not general conditions? 'General conditions' is a generic term, which has different meanings to different parties in different situations, but it is still important that it be used properly. GCs are general, in that they do not have a specific focus. In the case of the home office, the GCs funds the financing of the operations of the company, which is outside of and not applicable to individual jobsite operations. Essentially if there were not any construction projects, the company would still have GCs activities and costs, but it then would not be in business for very long. Similar to the list presented in Chapter 5, some alternate terms or use of the phrase GCs include:

- American Institute of Architects (AIA) A201 is a general conditions contract attachment to most standard AIA contracts;
- Construction Specifications Institute (CSI) divisions 00 and 01 include general conditions but more often designated supplemental or special conditions to the contract;
- Overhead or overhead costs or general overhead;
- Indirect costs, in that they are not direct to the cost of construction;
- General requirements;
- General and administrative costs often are how general conditions or overhead is described by public entities;
- Distributable costs is another name for general conditions, as overhead is distributed over the costs of the entire company operations and not attributed to any particular project or construction activity;
- Administration or administrative or operational costs;
- Overhead and profit is known as fee for contractors. If the contractor anticipated 2% home office general conditions cost and desired a 3% profit, they would bid a project with a 5% fee; and
- Overhead for consulting companies includes the cost of the work plus profit. There is not any profit factored into overhead for contractors.

This chapter focuses on *home office general conditions* and not jobsite general conditions. It can be difficult for the built environment novice to navigate through all of these terms which in some cases are similar and others quite different. One easy way to keep them separate is to insert the terms jobsite and home office as a prefix, such that *jobsite administrative costs* and *home office GCs* are easy ways to keep them separate. Many participants in the built environment use the abbreviation GC for general conditions as well as for general contractor.

There are two ways to prepare a general conditions estimate. The first is very easy, and that is to simply apply a pre-determined percentage to the anticipated yearly volume produced by the construction operations. For example, assume the corporate officers anticipate on January 1, 2020 that the company will have a volume of $100 million in 2020, based on their current backlog (projects under contract but not yet completely billed), and projects which are being pursued. If last year's home office GCs costs were $2 million and this year was expected to be similar, this year's construction projects would need to generate a total fee of $2 million to break even. This equates to 2% of $100 million in total expected revenue. Each project would need to produce a 2% fee for the company to, at a minimum, just break even. This can all be shown by the following equations. The problem with estimating with a rough percentage, such as this, is that it does not consider changes from one year to the next.

> Total anticipated revenue: $100,000,000
> Total anticipated home office overhead: $2,000,000
> General overhead percentage: $2,000,000 / $100,000,000 = 2%
> Minimum fee each project must generate to break even: 2%

The more accurate way to prepare a home office general conditions estimate is to prepare a detailed line-item estimate, exactly as was done for the jobsite GCs in the last chapter. There are home office administrative labor and personnel categories, materials and supplies, and office equipment. This is all very similar to the jobsite estimate. Each line item likely has a quantity or duration, such as 12 months, or 52 weeks, which are multiplied times labor rates and material unit prices. An abbreviated home office GCs budget is included as Figure 6.1.

The complete estimate may be 100 to 200 line items long, depending upon the size of the company and the proportion of management which is performed at the home office compared to the jobsite. The general overhead budget should be detailed enough to track, but not so detailed that it is unmanageable, similar to the 80-20 rule that is used throughout this discussion of cost accounting and financial management. Different categories of office labor should be kept separate and not combined as they all have different rates. The process for estimating the home office general conditions would be similar to estimating direct construction costs. Historical costs are available, trends are analyzed, and new conditions of the coming year must be considered. At the completion of this year, the actual costs should be fed back into the GCs database to better prepare future estimates.

There are many potential home office 'staff' personnel who may or may not be billable to the project. In the case of a lump sum project, it doesn't matter as it is closed-book and the client does not really care how costs are accounted, unless they show up on change order proposals. In a

Evergreen Construction Company						Date:		01/01/2020
Annual Home Office General Conditions Budget, Abbreviated						Estimator:		Robert Benson, OIC
						Estimate #:		1
Home Office General Conditions								
Line Item	Description	Qty	Unit	Wage Rate	Labor Cost	Unit M Cost	Material Cost	Total Cost
5	Receptionist	12	mos	$3,000	$36,000			$36,000
14	VP of Operations	12	mos	$10,000	$120,000			$120,000
56	Cleaning Service	52	wks			$800	$41,600	$41,600
72	Office Supplies	1	LS			$25,000	$25,000	$25,000
99	Copy Machine Rental	12	mos			$700	$8,400	$8,400
141	Warehouse and Storage	12	mos			$3,000	$36,000	$36,000
	Continued…							
	Totals:				$1,150,000		$850,000	**$2,000,000**

Figure 6.1 Home office general conditions budget

negotiated project, some home office personnel and costs may be attributable to a project, but that depends upon the contract terms. The AIA A102, cost plus fee with a guaranteed maximum price (GMP), is the contract that Evergreen Construction Company (ECC) executed on the Olympic Hotel and Resort project. Contract Article 7 in that agreement spells out the cost-reimbursable items and Article 8 lists those which are non-reimbursable. Generally any cost incurred at the site is considered a reimbursable cost and any function carried out at the home office is considered non-reimbursable. Any additional home office general conditions item or person can be inserted into Article 7 as a reimbursable cost, but it must be mutually agreeable and done so before the contract is executed. Some of the home office GCs costs which can be inserted into the contract include:

- Scheduler,
- Estimator,
- Safety officer or inspector,
- Quality control officer or inspector,
- Data processing costs, including accounting and audits,
- Cost engineer or accountant,
- Senior project manager,
- Preconstruction services fee or costs,
- Specialty superintendents, such as a hoisting superintendent, concrete finisher superintendent, earthwork superintendent, ironworker or structural steel superintendent, and others.

It is very common to see the project manager inserted as a reimbursable cost in Article 7; this was the situation on the hotel case study project. The jobsite cost accountant was also inserted as cost-reimbursable and her wages were included in the jobsite administration estimate. An example of an expanded annual home office general conditions estimate for Evergreen Construction Company can be found on the eResource.

Management of the home office general conditions budget

Management of the home office general conditions budget is the responsibility of the CFO and CEO and is not project-based. The contractor's goal is to reduce its home office overhead costs with respect to overall volume. Recall that fee equals overhead and profit. If the fee is fixed and home office costs are reduced, then profit is increased as reflected in the following equations. The jobsite team focuses on producing fee, and the home office team uses that fee to first pay its operational costs. Anything remaining is considered profit.

$$\text{Fee} = \text{Overhead} + \text{Profit}$$
$$\downarrow \text{Overhead} = \uparrow \text{Profit}$$

The lack of proper understanding, recognition, management, and allocation of home office overhead to construction projects is a common source of smaller-sized contractor failures; an example is exhibited in the following. The goal of HOOH management is first to make sure the company spends within its budget, just as they would do with jobsite indirect and direct costs. One way to do this is not to add additional overhead expenditures, such as a second marketing director or new office equipment, beyond that which was anticipated on the first of the year and accounted for in the annual budget. The second goal is to reduce overhead costs, and that can be accomplished in a couple of fashions:

- Reduce overhead costs by eliminating personnel. It is unlikely that wage rates can be reduced, but if one accountant or one receptionist or one vice president leaves during the middle of the year, and is not replaced, costs will be reduced. This applies to materials and office equipment as well.
- Attribute as much of the home office overhead to construction projects as is possible. Examples include wages of the project manager and jobsite cost accountant as discussed earlier.
- If construction equipment is owned by the construction company, and not a separate equipment company, it needs to be out on jobsites and charged to projects and not in the warehouse or the company storage yard. Equipment maintenance also needs to be job-costed wherever possible as will be expanded on in Chapter 12, 'Equipment use and depreciation.'
- Labor burden is a combination of labor taxes and labor benefits. All labor burden, including burden on direct and indirect project labor, should be attributable to the project and job-costed and not costed to the home office. Labor burden is included with Chapter 18, 'Taxes and audits.'
- Overhead is a combination of fixed and variable costs as discussed with breakeven analysis later. If a company's volume is increased, the variable overhead costs increase proportionately, but the fixed overhead costs are not necessarily increased. If revenue is increased just to the point, but not over, where a jump in fixed overhead costs would be necessary, such as the addition of an administrative clerk or a second payroll clerk, then the total overhead percentage as compared to total revenue is reduced.

Example One: This small but successful custom and spec home builder was organized as a sole-proprietor and utilized the cash accounting method of recognizing revenue and expenditures. The owner was also the project manager, superintendent, lead carpenter, and on weekends estimated new projects. His wife was the bookkeeper, interior decorator, assisted with marketing, and often was seen picking up materials in her husband's pickup truck. He invoiced his clients his time at $2 per hour above union scale, his crew's actual wages straight through, and marked up materials and subcontractors by 10%. He did not factor his home office or equipment utilized in his home office into his billings. Neither did he invoice for his wife's wages, his pickup truck, or his shop/warehouse. He made a decent and fair income, due to an outstanding relationship in the community for quality work and fairness, but when it came time to replace his pickup truck and radial-arm saw, he had to dig into his personal savings.

Any reduction in home office overhead, assuming a fixed fee, increases profit. But a reduction in the overall overhead percentage as related to revenue also allows the contractor to reduce its required fee and become more competitive with their bids and proposals. In the previous example,

HOOH was set at 2% and profit at 3% with a desired fee of 5%. If HOOH were reduced to 1.5% through any of the means discussed earlier, and the contractor retains their goal of a 3% profit, then the fee (which could either be used for bids or open-book proposals) could be reduced to 4.5%. The lower fee should then increase revenue which results in additional opportunities to make additional fee with more projects.

Some smaller contractors will attribute more of their expenses to the home office than do others, simply because it is easy. These include many percentage markups such as liability insurance and labor burden, even for craftsmen. In this case, personnel who work on many projects, such as a project manager running five jobs, a structural steel specialty superintendent who helps all the projects with steel, or a quality control inspector who visits ten jobs a week for a half a day each, are all cost-coded to the home office. In the contractor's view, 'cost is cost.' But ideally each project should carry its own share, such that the firm can easily disseminate which projects are more successful. There are several advantages to job-costing versus home office costing on negotiated projects as well as is discussed throughout this book.

There are a couple of different ways home office overhead may be allocated to projects. The most common is for it to be distributed proportionately across all of the projects in the company. If one project accounts for 50% of this year's corporate volume, than that job needs to generate enough fee to cover 50% of the HOOH expenses. In this way every project carries its own weight. Other methods include allocating more of the general conditions to projects which have a higher fee opportunity; allocate them based on project durations – those that last longer are charged more; or attempt to subjectively allocate them based on project support needs. Conversely if a company has only one construction project, that project must pay for 100% of the home office costs just to break even. Another option is to reorganize the company and/or office as exhibited in Example Two. The application of overhead costs to divisions, projects, and direct construction activities is the basis behind activity-based costing (ABC) which will be discussed in detail in Chapter 10.

Example Two: This commercial general contractor had followed a repeat negotiated client to a new city and opened a branch office with expectations of additional work. The additional work did not materialize, for a variety of reasons, and the contractor had only one competitive bid project in backlog for the next year. They shut down their nicely appointed rental office, laid-off most of the office personnel, and the branch manager and his receptionist moved to the jobsite trailer and assumed the roles of project manager and administrative assistant/cost engineer. The PM which had previously been assigned to that job stepped down into the project engineering (PE) role. They survived that year, expanded into a permanent office the following year, and are now the largest general contractor in that city.

Breakeven analysis

Construction projects are required to bring in a sufficient fee to cover home office operations costs. If the home office general conditions budget is $4 million then $4 million in fee is a minimum the company needs just to pay its bills. This fee is the breakeven point, with zero profit, which is below

Table 6.1 Breakeven analysis

Ideal Revenue	Breakeven Revenue Without Profit	Construction Cost	Stepped Fixed OH	Variable OH %	Variable HOOH	Total HOOH	Total HOOH % of Cost	Ideal Profit as % of Cost	Fee, or OH&P	Ideal Net (Before Tax) Profit
$50,000	$50,000	$0	$50,000	2.0%	$0	$50,000	NA	4.0%	NA	$0
$156,000	$152,000	$100,000	$50,000	2.0%	$2,000	$52,000	52.0%	4.0%	56.0%	$4,000
$630,000	$610,000	$500,000	$100,000	2.0%	$10,000	$110,000	22.0%	4.0%	26.0%	$20,000
$1,160,000	$1,120,000	$1,000,000	$100,000	2.0%	$20,000	$120,000	12.0%	4.0%	16.0%	$40,000
$2,220,000	$2,140,000	$2,000,000	$100,000	2.0%	$40,000	$140,000	7.0%	4.0%	11.0%	$80,000
$10,850,000	$10,450,000	$10,000,000	$250,000	2.0%	$200,000	$450,000	4.5%	4.0%	8.5%	$400,000
$53,650,000	$51,650,000	$50,000,000	$650,000	2.0%	$1,000,000	$1,650,000	3.3%	4.0%	7.3%	$2,000,000
$107,000,000	$103,000,000	$100,000,000	$1,000,000	2.0%	$2,000,000	$3,000,000	3.0%	4.0%	7.0%	$4,000,000
$532,000,000	$512,000,000	$500,000,000	$2,000,000	2.0%	$10,000,000	$12,000,000	2.4%	4.0%	6.4%	$20,000,000
$1,063,000,000	$1,023,000,000	$1,000,000,000	$3,000,000	2.0%	$20,000,000	$23,000,000	2.3%	4.0%	6.3%	$40,000,000

Volume = Revenue = Contracted Construction Cost (Direct + Indirect) + OH&P (Home Office Overhead (Fixed + Variable) + Profit) where the corporate equity owners desire it to be. They are expecting an above average return on their investment because of the high risk investment they made in a construction company.

Home office and jobsite general conditions costs both can be thought of as a mix of fixed and variable costs. Fixed overhead costs are those which are more time-dependent than volume- or revenue-dependent. The office rental for the company costs $400,000 per year. If the company accomplishes either $1 million in volume or $500 million in volume the office rent is the same, it is fixed. These fixed costs may increase though when volume has increased so substantially that an addition is necessary. This may be the case with added office space or an additional company officer or accountant. It is difficult for a contractor to reduce its fixed overhead in an effort to increase its profit margin.

Variable overhead costs are those that are volume-dependent. Items that would show up at the bottom of a contractor's estimate summary page, those that are below-the-line and are percentage add-ons, are variable costs. These include items such as liability insurance, excise tax, and small tools, or in the case of home office overhead, office supplies. If the contractor has a year with very little business, then although they still may need a copy machine (fixed cost), they do not need to run as much paper through it (variable cost).

Evergreen Construction Company is anticipating a volume of $100 million in 2020. The *fixed home office general conditions* estimate is a series of line items which is mostly time-dependent and is anticipated to cost $1.5 million this year. The *variable general conditions* costs are volume related and they are budgeted at 0.5% for the coming year. $100 million multiplied by 0.5% is an additional $500,000 in general conditions, for a total of $2 million. Table 6.1 is a sample breakeven analysis worksheet which combines revenue with fixed and variable overhead costs, construction costs, and fees.

Profits

Contractors set many goals including building quality projects, meeting schedules, keeping everyone safe, and developing a good reputation with clients and subcontractors. But contractors are also in the business of making a profit. Construction is risky and is not a 'not-for-profit' industry. Project managers will be reminded of their fee goals throughout the course of construction on their projects by the home office. One of the unique aspects of construction is that true fee will not be known until the project is complete. In the home office, corporate executives will not know what the total yearly profit will be until the year is complete and all job costs and revenues have been factored, along with the actual home office general conditions expenditures. This section analyzes methods to determine estimated profit, sources of profit, and methods to improve profit.

Methods to determine estimated profit

Profits are not what contractors add to the bottom of construction estimates, they add a proposed fee, but for some participants in the built environment, they see the fee as all profit. The fee is also known as the 'margin' or generically the 'markup.' But as we have discussed earlier, the fee first needs to cover home office overhead, and any money remaining after overhead is accounted for may be considered gross profit. The following accounting equations will help explain how net profit is derived from the original contract value. It is the net profit that provides a return on equity (ROE) for the company owners. Net profit is also known as after-tax profit or net income or 'the bottom line.' Revenue would be the top line in an income statement as reflected in the next chapter's discussion of financial statements. Then after a series of expenditure deductions, and subtotals, a summary of which is reflected in the following equations, the net profit is reflected on the bottom line.

Revenue, or Contract value or Volume = Total cost + Gross profit
Total cost = Construction cost + Home office overhead cost
Revenue = Construction cost + Fee
Construction cost = Direct construction cost + Jobsite administration cost
Direct costs = Direct labor + Material + Equipment + Subcontractors
Fee = Home office overhead + Gross profit, or OH&P
Net profit = Gross profit – Income taxes

When a contractor is preparing to bid a project, or submit a proposal on a negotiated project, they have to determine what fee to include. Some of the fee considerations the contractor's CEO will make include:

- The fee must at least cover the home office breakeven overhead costs, which are approximately 2–3% of revenue for a mid-sized commercial general contractor.
- Direct labor is the biggest estimating risk for any contractor. The fee should cover approximately 50% of estimated direct labor cost for a commercial general contractor. This is saying that if the estimator missed the labor estimate by half, they will still come out even.
- There is an opportunity cost for both the project manager and superintendent working on any particular project. The PM and superintendent are soft assets with earning power. They are expected to earn a fee for their company. If they cannot realize a fee of $20,000 per month each on project 'A' then the contractor should pass on it and pursue project 'B'. An experienced project engineer or cost engineer can be added to this equation as well.
- Does the contractor have a large backlog, in which case the fee would increase, or does the contractor need additional backlog, in which case the fee would drop?
- Market conditions may indicate that work is being bid today with a fee of 4–6%, therefore the fee needs to be within that range.
- Availability of personnel, especially the PM and superintendent.
- The contract type can significantly influence the fee decision. Lump sum projects have increased risk for the contractor and deserve more fee than a cost-plus project. A project with a guaranteed maximum price is somewhere in between.
- There are many contract issues and contract clauses which can raise or lower the fee, such as liquidated damages (LDs) and definition of reimbursable costs in the AIA A102 contract, Articles 7 and 8. If there are additional opportunities for home office costs to be considered job cost and therefore reimbursable, this allows money to be moved from the home office general conditions to jobsite GCs and allows a lower fee, all the while retaining the same profit goal.

Other potential risks or opportunities such as the client and subcontractors and design team can have an influence on increased or reduced fees as well. The greater the risk, the higher is the fee. If the risks are too great, there might not be any fee that would make the job attractive. Risks are also resolved by the contractor purchasing insurance, bonding subcontractors, increasing estimating contingencies, or taking on a joint venture partner for a specific project.

Sources of profits

There are many ways a contractor can make money and realize a profit, just as there are many ways they can lose money. The following are some examples.

- Operate each division within the company and each separate construction project as independent profit centers. This is one focus of activity-based costing.
- Minimize home office overhead as discussed earlier. This is difficult to do and is not a financial management responsibility of the project manager.
- Perform more work with direct craft labor. If direct-hire carpenters and ironworkers beat their estimate, then the general contractor makes additional fee. But as will be discussed in Chapter 8, 'Cost control,' this also increases risk.
- Alternatively, reduce direct labor and increase the mix of subcontractors. This will not necessarily improve profit potential, but it reduces risk and therefore improves the potential to make the estimated fee.
- Individual construction crews or teams, including foremen, and superintendents are potential profit centers as some may consistently beat the budget.
- Build the project quicker: saving times saves jobsite general conditions expenses.
- Build with good quality and minimize rework, which is a goal of lean construction.
- Construction accidents cost everyone in a variety of fashions. Safer projects have proven to be more cost efficient.
- The project management team does not have as immediate an effect on profits as does the construction crew, but they need to be held accountable to return the estimated fee if not improved upon. They can do this if they are given the authority to act and make independent decisions in the field and are not totally reliant on home office oversight.
- Reduce construction costs by:
  ◦ Beating labor productivity estimates,
  ◦ Efficient material purchases and deliveries with less waste,
  ◦ Shorter construction schedules, and
  ◦ Reduced jobsite GCs costs.
- Efficient bid procedures and later successful buyout of subcontractors and suppliers and execution of tight contracts.
- Types of projects: some companies will make more money on retail and others on hotels.
- Clients or project owners can make a project run smoothly and improve a contractor's efficiency which should prove it more cost effective. The reverse is true as well. Client satisfaction is the responsibility of the PM and superintendent, not only on this project, but for future references. Some projects have potential incentive fees calculated from scorecards to be completed

by the client which provide the general contractor with a bonus based on several categories of performance, including schedule, quality, safety, and communications.

- Efficient use of construction equipment, including:
  - Economical rental periods, such as one week and not four days,
  - Good business decisions either to rent from external firms, or require subcontractors to provide all the equipment, or rent from in-house equipment companies, and
  - Open-book contract issues affecting equipment maintenance repair costs and rental rates.
- Some geographical locations, including cities and states may be better for one contractor versus another due primarily to relations with subcontractors.
- Some architects and engineers are easier to work with than others which can have a positive impact on cost performance.
- Guaranteed maximum price contract savings splits.
- Back charging subcontractors for work which was either performed by the general contractor or another subcontractor or for work which the general contractor had to repair.
- Sliding scale fees allow a contractor to make more money on direct work than on subcontracted work and more money on change orders than on the base contract.
- Some below-the-line markups have the potential for hidden fees, in that the markup charged either on change order work or on the original open-book contract was higher than it needed to be. Some of these can be discovered during a financial close-out audit, but others are very complicated. This includes markups such as labor burden, liability insurance, data processing, and others.
- And for some contractors, but unfortunately for owners and architects, claims are a source of potential profit.

Methods to improve profits

The contractor's goal is first to achieve the fee that was estimated, and then improve upon that. Assuming the home office overhead costs are set, an improvement in the fee directly results in an increase in profit. But without systems in place to first estimate accurately and then track costs, as shown in the following example, profits cannot be attained, let alone increased. There are many ways that *profits can be improved*, including:

- Raise bid prices, but this may also reduce volume;
- Increase volume without increasing fixed overhead;
- Reduce fixed overhead expenditures, with a constant volume;
- Specialize in one type of construction, for example medical facilities;
- Be selective with client choices, types of work, designers, subcontractors, employees;
- Improve preconstruction planning;
- Reduce construction cost; and/or
- Learn better methods. See Chapter 11 'Lean construction techniques.'

Example Three: This small speculative home builder did not have any means of construction cost control and did not even track the cost of any individual home. He often had three houses under construction at any point in time, one at the site-clearing stage, one in rough framing, and the third in trimming-out interior finishes. The economy was slow, but he was still able to sell a couple of houses each year and pay his bills. There were two things he did not understand: first, he was using the revenue from the current house sale to pay for the bills on the last house, and second, he was selling them below cost. When the economy slowed even further, and he was stuck with two completed but unsold houses, he went through bankruptcy and lost the houses to the bank.

Contractors may choose to *raise their estimated or bid fee* on one particular project for a variety of reasons, including:

- Provide a practice bid, although this is expensive to do and potentially dangerous if the contractor accidently becomes the low bidder;
- Market conditions allow a higher fee;
- The type of work is not the contractor's specialty;
- The client may be a high risk client which is prone to slow payments or lawsuits;
- The construction documents are not complete, although some contractors would see this as a change order opportunity;
- Tight schedule mandated by the client, possibly with liquidated damages;
- Other risky contract clauses, such as slow payment terms or high retention withheld;
- Provide a courtesy bid to a client they do not want to offend if they have a sufficient backlog; and/or
- If the project is difficult or complicated it may require a higher fee due to higher associated risks.

Other markups and add-ons to the estimate

The fee is only one markup that is placed *below-the-line* in the estimate summary; there are several others. All of the estimated costs 'above-the-line' are considered direct costs. Similar to fee, other items below-the-line are percentage add-ons to the estimate and are volume-dependent and therefore are variable costs. Labor burden is applied to direct and indirect labor only, not to the entire direct cost subtotal. Labor burden is not applied to material costs and subcontractors are expected to have covered their own labor burden in their bid prices. Different burden rates should be applied to direct and indirect labor. Typical percentage add-ons that will show up below-the-line on a construction summary estimate include: labor burden, liability insurance, excise tax, sales tax, contingency, and fee.

Summary

The home office general conditions estimate is created and managed out of the home office by the CFO and CEO. The jobsite GCs estimate, as discussed in our previous chapter, is developed by a staff estimator and/or the project manager. Different than the home office GCs, the jobsite GCs are locally managed by the jobsite team, including the PM, superintendent, and jobsite cost accountant. It is important that the jobsite team have an understanding of what is included in the home office GCs, or overhead, as overhead plus profit equals fee. The jobsite needs to focus on the fee and return at least what was estimated in order to cover home office general overhead and return a fair profit to the construction company owners.

The creation of the home office general conditions estimate is similar to other estimates in that it is comprised of several line items, including labor and material and office rent and equipment. The quantity or duration for each of those items is multiplied by the relevant wage rate or unit price. The home office will look for various ways to reduce its overhead costs, one of which is to move home office activities out to the jobsite where, on cost-reimbursable projects, they might be able to be paid out of job cost, assuming the contract language allows this. There are many ways to improve profit potential for a construction company, in addition to the reduction of home office GCs. As a close project manager-friend once told this author, "you have to keep your eye on the prize," and the prize is the fee, which includes profit.

Review questions

1. Why do we even bother with a discussion of HOOH when our discussion in this book is focused on jobsite financial management issues?
2. List three different uses for the term 'general conditions.'
3. Is HOOH billable to the client on a) lump sum jobs, and/or b) open-book negotiated jobs?
4. How does the contractor find out what home office personnel may be charged to a project, and when should this be investigated?
5. If the small spec home builder in Example Three had changed his volume to a) just one home per year, or conversely b) to 20 homes per year, would he have been more successful and been able to make a profit?
6. Looking back to the breakeven Table 6.1, how can the contractor increase its profit percentage by reducing volume?
7. You may need to look back to Chapters 3 and 4 for this question. Compare the profit potential on three different contracting methods – lump sum, cost plus percentage fee or time and materials, and GMP.
   a. On which type of contract can a contractor realize the highest fee?
   b. On which type of contract can a contractor lose the most money?
   c. Which of the three contracts is the only one which guarantees the contractor some fee?

Exercises

1. Many reasons why a contractor might choose to increase its bid fee were discussed in this chapter. Why might a contractor choose to lower a lump sum bid fee or negotiated proposal fee?
2. When should the PM be located at the home office or the jobsite and is he or she then billable?
3. Without looking at the detailed home office GCs estimate on the eResource, what other items might be included beyond those which were listed in Figure 6.1?
4. What happens if you as a PM do not make your bid fee a) on one project, or b) on all of your projects?
5. What happens to the construction company if it doesn't make a profit a) for one year, or b) for several years in a row?
6. How might construction costs be reduced?
7. Safety is not a topic in a discussion of cost accounting and financial management, but maybe it should be. How can an unsafe construction project cost more money?
8. If your CEO suddenly hired several new home office employees such as an additional marketing director, assistant payroll clerk, and a vice president in charge of estimating, what is the short-term impact on you and your project? What is the short-term effect on net profit? What is the long-term effect on bid and negotiated project fees?
9. If a contractor suspects a project might be risky to pursue, they have several options available to them. What might some of those be?
10. Prepare a net profit equation starting with revenue or construction contract, and apply a series of deductions, including all of the equation variables presented in this chapter, until net profit is the result.
11. Review the expanded home office GCs estimate for ECC from the eResource and identify five line items which are examples of fixed overhead and five which are examples of variable overhead. Do not repeat any of the examples discussed in this chapter.
12. What is the minimum fee, in both dollars and percentage of construction contract, a general contractor would need on an upcoming $10 million bid project? Assume the following parameters:
    - $100 million expected yearly corporate volume,
    - 2% HOOH budget,
    - Owner's equity is $4 million,
    - This project is estimated to have $1 million in direct craft labor,
    - This project is expected to last 12 months,
    - There is a full-time PM and superintendent without a PE,
    - Cost accounting is performed out of the home office,
    - Equity owners expect a 15% ROE.
13. How would your answer change in Exercise 12 if:
    a. Market fees are at 7%,
    b. This was a negotiated project,
    c. The GC will perform $1.5 million in direct labor, or
    d. The GC has sufficient backlog and is submitting a complementary bid to a past client?
14. Resolve the following breakeven questions using Table 6.1 There are an unlimited amount of questions and exercises that can be based on this table and analysis:
    a. What is the breakeven revenue based on $75 million of construction cost?

b. Based on a revised negotiated profit of 5% and $75 million in construction cost, what is the total revenue?

c. If a small contractor had an annual volume of $5 million and $4.5 million of that was construction cost, what would be its profit margin in percentage?

d. If a large contractor had an annual volume of $532 million, and it desired a net pre-tax profit of 5%, how much fee in percentage and construction cost would it need to support that goal?

e. Assuming a contractor could lock in on a fee (any % would do), other than cutting construction cost, and using only information you can garner from this table, how can it maximize its net profit percentage?

15. Given the information provided here about ECC's annual revenue and overhead expenses, how do they compare with the breakeven analysis of the contractor in Table 6.1?

7

Financial statements

Introduction

The purpose of financial statements is for the contractor to connect and communicate with all of its internal and external financial stakeholders, such as the contractor's bank, bonding company or surety, chief financial officer (CFO), other company officers, board of directors, and investors or equity partners.

There are three major ledgers used in construction cost accounting which include the general ledger, job cost ledger, and the equipment ledger. The general ledger is comprised of two important financial statements: the balance sheet and the income statement. Students of cost accounting and financial management and new project engineers (PEs) and project managers (PMs) will not be expected to create or even manage these documents. Management of corporate financial statements is definitely a senior management operation performed by the CFO, but the PM should know the differences between some of these documents and how they relate to jobsite financial management.

In addition to discussion of these financial statements, this chapter will also introduce some of the financial ratios which are created by comparing different line items and different reports. Journal entries are internal accounting operations which move money around, on paper (or in this case, in the computer), but do not affect the bottom line. There are other financial reports generated by the home office accounting department that are project related and utilized for cost control efforts that will also be discussed in upcoming chapters. All of these documents and systems assist the contractor with its short-term and long-term financial decision-making processes.

Balance sheet

Like so many of these financial reports, the general ledger is for the entire company and is not job-based. The general ledger is comprised of two important financial statements, the balance sheet which is discussed in this section and the income statement which will be discussed in the next section. The general ledger utilizes a chart of accounts, which is similar to cost codes used in the original construction estimate and for cost control, but is slightly different from jobsite codes as it is not based on the Construction Specifications Institute divisions. The chart of accounts for the general ledger is usually just a three-digit code as reflected in the following list, but there are variations between factions of the construction industry and individual contractors. Similar to cost codes, there are not any rules per se as to exactly which numbers must be used. Our construction company certified public accountant (CPA) utilizes the following code system:

- 100–149: Current assets
- 150–199: Fixed assets
- 200–299: Liabilities
- 300–399: Net worth or owner's equity
- 400–499: Revenue
- 500–599: Jobsite expenditures
- 600–699: Home office expenses
- 700–799: Taxes

An abbreviated example balance sheet for Evergreen Construction Company (ECC) is reflected in Table 7.1. The balance sheet represents the company's net worth. This is a snap shot in time and answers the basic question that so many of the stakeholders mentioned earlier ask: "How much are we worth?" The balance sheet is dynamic, it is always changing and in a constant state of flux. It is a snapshot statement of a contractor's financial position. Every check written, and every check received, immediately affects the balance sheet. The balance sheet has three major sections, which comprise general ledger codes 100–399, and include: Assets, Liabilities, and Owner's equity, or Net worth.

Assets represent positive entries on the balance sheet and liabilities are negative. Assets reflect what a contractor 'has' and liabilities reflect what the contractor 'owes.' The total of all of the assets does not answer the 'worth' question earlier, as liabilities need to be deducted from the asset total. Worth is defined as what is left over, or in this case, the owner's equity. The owner's equity does not necessarily represent what was originally invested, but rather what that investment is worth at any given point of time. The key to the balance sheet is it must 'balance.' The sum of the assets will always exactly equal the sum of the liabilities plus owner's equity. If there is an increase on one side of the ledger, there must be a corresponding increase on the other. The balance sheet equation is reflected as:

$$\text{Assets} = \text{Liabilities} + \text{Owner's equity, or}$$
$$\text{Assets} - \text{Liabilities} = \text{Owner's equity}$$

Table 7.1 Balance sheet

Evergreen Construction Company
Balance Sheet
Statement Date: December 31, 2020

Code	Category	Values
	Assets:	
	Current Assets:	
110	Cash	$250,000
120	Accounts Receivable	$9,575,000
130	Inventory	$0
140	Other Current Assets	$10,000
	Total Current Assets:	$9,835,000
	Fixed Assets:	
150	Land	$1,000,000
160	Buildings	$2,800,000
170	Construction Equipment	$435,450
180	Office Equipment	$75,677
190	Accumulated Depreciation	-$1,165,050
199	Other Fixed Assets	$12,550
	Total Fixed Assets:	$3,158,627
	Total Assets:	$12,993,627
	Liabilities:	
	Current Liabilities:	
210	Accounts Payable	$5,550,000
230	Notes Payable	$75,000
240	Accrued Expenses/Taxes	$85,000
270	Other Current Liabilities	$15,550
	Total Current Liabilities:	$5,725,550
	Long-Term Liabilities:	
289	Long-Term Debt	$1,000,000
290	Other Long-Term Liabilities	$18,077
	Total Long-Term Liabilities:	$1,018,077
	Total Liabilities:	$6,743,627
300	Owner's Equity	$5,750,000
	Retained Earnings	$500,000
	Total Liabilities and Owner's Equity:	$12,993,627

Assets are categorized either as *current assets* (also known as short-term assets), versus *fixed assets* (also known as long-term assets). Current assets are those held for less than a year and include primarily cash or accounts receivables. Receivables for a construction company are defined as invoices processed but not yet received from its clients. Current assets are also considered liquid assets in that they can be converted to cash within a year or less. Fixed or long-term assets are those which are held for a year or longer and include buildings, furniture, and construction equipment. As will be discussed in upcoming Chapter 12, 'Equipment use and depreciation,' and Chapter 19, 'Developer's pro forma,' construction companies generally do not hold a substantial amount of fixed assets. Instead they organize separate independent companies which own and manage them as independent profit centers in addition to, but legally separate from, the construction company. These separate entities are known as limited liability companies (LLCs).

Liabilities are also split between *current* and *long-term liabilities*. The comparisons are very similar to assets, except they reflect negatively on the balance sheet instead of positively. Examples of current or short-term liabilities are those owed or payable within one year and classified as accounts payable. Accounts payable for a contractor includes money invoiced by subcontractors or suppliers but not yet paid, and time incurred by, and wages owed to, the contractor's employees. Loan payments owed on buildings and equipment are long-term liabilities.

The remainder, after liabilities are subtracted from assets, is the *owner's equity*, which is also known as net worth or stockholders' equity. The original owner's equity was the initial investment, in the form of cash or capital, made into the company by one or more partners. The current owner's equity reflects the value of the investment as of the date of the balance statement. The owner's equity is shown among the liabilities on the balance sheet as the company essentially owes that money to the owners. If the company would dissolve at any given point of time, the assets would be sold and used to pay off the liabilities and the remainder would be owed and distributed to the owners. There would not be anything left over after that; the assets and liabilities effectively balance out.

During a good year, after the owners have taken their dividends out of the after-tax net profit, some of the profit may remain in the company as retained earnings but still carried within the owner's equity portion of the balance sheet. Retained earnings can be used in the future in a variety of fashions including: company expansion into a new market or with additional office space, equipment purchases, or carry the retained earnings forward in case next year's profits are not as good as was this year's.

Income statement

The income statement is a sum total of all of the business that the company did for a period of time, usually one fiscal year. It can also be prepared quarterly or monthly. The income statement combines revenue and cost and is also known as the profit and loss statement. The income statement reflects the differences between two balance sheets which were both snapshots in time. General ledger chart of accounts codes 400–699 are covered in this statement. Revenue is the money which came into the company or total volume of business. This is how much was 'earned.' Cost is the money which left the company, in the form of expenditures, or how much was 'spent.' The income statement does not reflect cash flow for most contractors, other than those small firms which use the cash accounting method, but rather committed costs and earned income. The income statement will not balance out to zero as the balance sheet did, at least the contractor hopes it doesn't sum to zero. Rather the income statement should reflect how much profit the company made for the year or period reported. The income statement equation is reflected as:

$$\text{Revenue} = \text{Cost} + \text{Profit, or}$$
$$\text{Revenue} - \text{Cost} = \text{Profit}$$

Table 7.2 is an abbreviated example income statement for Evergreen Construction Company for the year ending December 31, 2019. The company had five projects that year and all are reflected on this one statement, including the totals. It is likely the contractor would share only the far right hand total column with many of its external stakeholders.

There are several sub equations which are derived from the revenue equation, which include:

$$\text{Cost} = \text{Direct construction cost} + \text{Jobsite indirect cost} + \text{Home office overhead cost}$$
$$\text{Direct construction cost} = \text{Labor} + \text{Material} + \text{Equipment} + \text{Subcontractors}$$
$$\text{Fee} = \text{Home office overhead} + \text{Profit, or OH\&P}$$
$$\text{Net profit} = \text{Gross profit} - \text{Taxes}$$

Many of these equations are used in different chapters throughout this study on cost accounting and financial management. Some jobsite construction teams feel that their contract amount is how much they contributed to the bottom line; they built a $24.5 million hotel therefore they contributed $24.5 million. But as reflected by the series of formulas earlier – starting with revenue, or contract value, of $24.5 million – the amount which is left, after all costs are deducted, which contributes to the owner's bottom line in the form of additional owner's equity as discussed with balance sheet earlier, is very small. Even the profit is not pure profit. Gross profit, which is what is left after cost is deducted from revenue, is subject to taxes, which was changed for 2018 to a flat 21% for a construction company. After tax, profit is also known as net profit or pure profit.

Financial ratios

Financial ratios are a mathematical means to analyze all of the different ledgers and financial statements produced by and available to the company. These ratios and formulas provide an insight to the contractor's financial health beyond just the statements discussed earlier. An entry of $500,000 in any line on any of the statements does not tell the whole story without comparing that figure with other entries, possibly from other statements. Often ratio averages or trends are developed from two or more statements. There are many formulas available for bookkeepers, CFOs, banks, and CPAs. Financial ratios are also very important to real estate developers seeking investor contributions or bank loans. Different stakeholders have interest in different ratios. Construction project managers again will not be using these ratios, but it is important that they have a general understanding that a) these financial ratios exist, and b) financial transactions at the jobsite level have some impact at the corporate level. Most of the ratios are based upon line-item entries in the balance sheet and income statement. Some ratios compare items within one statement and others compare items between the two statements.

Table 7.2 Income statement

Evergreen Construction Company
Income Statement
Fiscal Year Ending December 31, 2020

Code	Category	Job Number:	109	114	332	410	422	2020
		Project Type:	Retail	Biotech	Medical	Spec/Shell	Olympic	Total
			T.I.	*Lab*	*Clinic*	*Office*	*Hotel*	*Total*
400	Revenue		$2,491,171	$18,698,787	$61,024,000	$5,000,010	$9,812,006	$97,025,974
	Jobsite Expenses:							
510	Direct Labor		$125,153	$3,110,050	$5,099,110	$51,060	$651,829	$9,037,202
520	Direct Material		$221,918	$3,202,777	$7,123,888	$95,023	$1,329,382	$11,972,988
530	Equipment Rental		$30,483	$362,800	$950,545	$101,009	$242,991	$1,687,828
540	Subcontractors		$1,768,411	$7,800,777	$38,937,870	$4,000,890	$5,889,881	$58,397,829
550	Indirect Labor		$64,500	$727,990	$2,194,051	$175,750	$353,712	$3,516,003
560	Indirect Material		$35,050	$406,005	$2,422,168	$235,789	$226,626	$3,325,638
570	Labor Burden		$84,669	$1,880,010	$1,389,000	$105,559	$429,715	$3,888,953
580	State and Local Taxes		$20,321	$143,316	$470,211	$45,679	$67,891	$747,418
590	Liability Insurance		$24,912	$182,846	$437,541	$43,213	$86,640	$775,152
	Total Project Costs:		$2,375,417	$17,816,571	$59,024,384	$4,853,972	$9,278,667	$93,349,011
	Project Fee		$115,754	$882,216	$1,999,616	$146,038	$533,339	$3,676,963
600	Home Office Overhead (see separate detailed spreadsheet)							$1,781,921
	Net Income:							$1,895,042
700	Income Taxes @ 21% Tax Rate							$397,959
	Net Income after Taxes:							$1,497,083

Notes:
All projects were started and completed in 2020 except job 422 which started in 2019 and completed in 2020.
Only the 2020 proportion of revenue and expenses for job 422 are shown in the 2020 income statement.

As introduced early in this book, construction is a major contributor to the built environment industries and construction has several different sectors or sub-divisions which primarily include: residential, commercial, heavy-civil, and industrial. The gauge of where any individual financial ratio should fall will be different depending upon what sector the contractor specializes in. The answer for a heavy-civil contractor is not the same as a residential contractor and general contractors will differ from subcontractors or specialty contractors. In addition, one year and one ratio are not true indicators as they reflect a narrow perspective and are just from a snapshot in time. Analyzing several ratios over years to see if there is a trend is more beneficial. Many contractor and accounting associations and publications report preferred ranges of many of these financial ratios, including the Association of General Contractors (www.agc.org), the National Association of Home Builders (www.nahb.org), *Engineering News Record* (www.enr.com), and Construction Financial Management Association (CFMA.org).

Financial ratios are typically grouped into categories. Four of the most common include: liquidity, profitability, efficiency, and leverage. It would be a lengthy treatise to cover every financial ratio that exists, as there is essentially an unlimited list. Following are a few of the more important ratios for a construction company's financial stakeholders:

Liquidity ratios provide a short-term analysis and indicate the company's ability to pay its bills:

- Current ratio = Current assets/Current liabilities
  The result must be greater than 1.3, ideally in the 1.5 to 3.0 range
- Quick ratio = Quick assets (cash or near cash)/Current liabilities
  The result should be in the 1.25 to 2.0 range

Profitability ratios: Many of these are analyzed both before and after taxes are deducted and reflect a longer-term analysis:

- Return on equity (ROE) = Both pre- and post-tax profit/Owner's equity (OE).
  Answers range considerably depending on the industry sector and the size of the company, but generally pre-tax ROE should be greater than 15%. This is one of the most important financial indicators for the company investors.
- Return on investment (ROI), same as ROE.
- Return on assets (ROA) = Net profit/Total assets, ideally greater than 5%.
- Net profit to Net worth ratio, ideally 10–20%, same as ROE.
- Net profit to Volume ratio = After tax profit/Construction revenue, ideally 1–3%.
- Earnings per share (EPS) = Profit/Quantity of shares.
  EPS applies to a corporation and not a sole proprietorship. The EPS and ROI may also be reflected on the bottom of the annual income statement.
- Rate of return (ROR) = Profit/Revenue, same as Gross profit ratio.
- This is the percentage of net income and can be project- or company-based.
- Gross profit or Fee percentage ratio = Gross profit/Construction revenue.

Efficiency indicators:

- Average age of accounts receivable measures the collection period from clients. Slow payments are generally bad, especially greater than 45 days. Receipt of payment by the 10th of the following month is the goal set out in Chapter 14, 'Payment requests.'
- Average age of accounts payable measures how fast contractors are able to pay their subcontractors. In this case, fast is good, ideally ten days after payment from the client is received and no later than 30 days.

Leverage indicators:

- Debt to Equity ratio = Total liabilities/Total OE, goal is 1.0 to 2.0.
- General overhead, or Home office overhead ratio = Overhead/Revenue. This is highly dependent upon a contractor's volume and how much of their business they are able to job-cost. Ideally less than 10%.
- Fixed assets to Net worth (or OE) ratio, ideally about 0.5.
- Backlog = Future revenue, not yet billed. Without a sufficient current backlog a contractor may be subject to suffering financial difficulties.

Journal entries

Financial transactions are *posted* to the *books* by cost accountants. Posting is defined as entering financial figures and the *books* are accounting statements or ledgers, some of which are discussed here. Journal entries are the method that cost accountants use to move money, by posting, at least on paper, between one account or cost code to another. The net effect of journal entries, without the addition of an approved change order, needs to be zero.

Some costs are entered into the job cost system by the home office accounting department, without processing invoices through the jobsite. These include items such as company-owned equipment, administrative labor such as project manager and superintendent, and each project's proportional share of markups to direct cost including home office overhead, labor burden, liability insurance, and excise taxes. Open-book clients occasionally struggle tracking home office journal entries during a final project audit. A cost shows up in March as one cost code and in August it is moved to another cost code. Or costs which were originally mis-coded to one project are transferred into another project. Journal entries are also performed by the jobsite team after they receive home office financial reports as discussed in the next section.

Job cost financial reports

Some of the reports generated by the home office accounting department, which are not company-related but are job-focused, include the monthly job cost history report, weekly labor reports, and an equipment ledger. There are many good and bad examples of cost accounting and financial management. It is said that more can be learned from our mistakes than from our successes. In this section there are three additional examples of contractor financial reporting that could have been improved upon.

Job cost history report

The job cost history report is also known as the job cost ledger. This report tracks the cumulative costs which have been coded to a specific job number. The report comes out about the 10th of the month which incorporates all of the costs from the previous month. This report is the sum effect of all of the labor time sheets, material invoices, and subcontractor invoice cost coding performed by the jobsite financial team. The home office journal entries will also be reflected in the monthly job cost history report. Each month the jobsite cost accountant will review the home office generated job cost history report to check that costs from other projects were not accidently posted to this project and that the cost codes he or she entered last month on time sheets and invoices were correctly input by the home office accounts payable clerk. There will always be cost coding errors, and it is imperative that they be corrected quickly to have a true picture of project costs to date. This will later help with the monthly cost forecast and the as-built estimate input. The cost codes entered on time sheets and material and subcontractor invoices should be the same codes as used in the original budget estimate, schedule codes, and file codes and should be consistent throughout the company. These are not the same as the general ledger chart of account codes introduced earlier. Coding errors must be corrected promptly utilizing the journal entry process.

> Example One: This large commercial general contractor had 20 project managers. The president of the company was very competitive and encouraged her employees to earn fees larger than the other PMs – she even posted the results for all to see. One unethical PM routinely would cost code his material invoices to other PMs' job numbers. After receipt of the monthly job cost history report, the other PMs would investigate these costs, if discovered, as many projects purchase similar materials from the same suppliers. If they found an incorrect entry, they would then try to journal entry it out of their job and into its rightful place. These types of unethical actions and confusing journal entry transactions make it difficult for certified public accountants to audit the books, especially on open-book negotiated projects.

Weekly labor reports

The weekly labor reports are a subset of the job cost history report but focus just on direct and indirect jobsite labor. The general contractor does not track subcontractor labor. Because labor is the most variable and difficult cost category for a contractor to estimate and control, labor reports are generated more often than material and subcontractor cost reports. They are typically produced by the home office and issued to the field, likely on Tuesday of the following week costs were incurred. This is to allow the jobsite management team to make corrections or modifications to the field cost control system, if necessary, as will be discussed in the next chapter.

Example Two: This cost engineer was responsible for tracking and reporting labor costs on a large hydroelectric project. Her focus area was for the electrician and plumber trades, which amounted to 600 and 300 employees respectively. At the end of every shift, foremen would write, in hand, often with poor hand writing, the work each individual craftsman performed that day, along with their hours. A separate timekeeper working for the payroll clerk would translate these time sheets into cost codes which would be entered into the accounts payable system each evening. The next morning the cost engineer would be presented with an Exceptions Report which would kick out all of the cost coding errors. She would spend the first half of each day tracking down the foreman and timekeeper and inspecting the work on the site to rectify these mis-codes and then journal entry them into correct codes. And the next day's activities would be a repeat of this one.

Equipment ledger

Contractors endeavor for all of their construction equipment to be job-costed and not charged or cost-coded to home office overhead. This includes company-owned equipment maintenance and repair expenses. Each piece of equipment is assigned a unique tracking number. A company equipment ledger may be created and managed by the home office accounting department. Heavy-civil contractors are more likely to own equipment than are commercial contractors. Many commercial contractors will form a separate company which owns and rents equipment to its jobs. These equipment companies are organized as partnerships or limited liability companies made up of the company executives. Taxes and liability issues of these companies are kept separate from the operations of the equipment company. But it is likely that one individual located in the home office accounting department is actually employed by the separate equipment company and manages the equipment ledger. Invoices may be sent out to the jobsite for cost coding by the site management team, or costs may simply be journal-entried into the job cost history report by the equipment clerk.

Example Three: This large residential contractor was also a developer and would build 1,000 homes each year on speculation of sale in several different tracts comprising 100 to 300 homes per tract. The four equity owners of the firm also owned a separate equipment rental company which rented equipment to the construction division. Each residential tract was set up as a separate LLC and had individual project managers and superintendents responsible for financial management affairs at the jobsite level. The company owners did not want any equipment sitting idle or cost-coded to home office overhead. The jobsite teams would routinely see equipment charged to their projects by the home office equipment clerk that was not necessarily on their project and were powerless to do anything about it.

Summary

The three major financial statements are the general ledger, which includes the balance sheet and the income statement, the job cost ledger, and the equipment ledger. The balance sheet by definition must balance. It is comprised of assets and liabilities. Owner's equity is included on the liability side of the balance sheet statement and is also known as the net worth of the company at any given point of time. The balance sheet is always changing. Anytime there is an increase on one side of the ledger there needs to be a corresponding increase on the other side. The income statement reflects the company's total revenue and total expenditures for a fiscal period such as one year. Whereas the balance sheet equals zero at the bottom, the bottom of the income statement, after deducting expenditures from revenue, is profit, and the construction company hopes to report a profit at the end of the year. Consistency in the preparation of financial statements is very important to internal and external stakeholders including certified public accountants, auditors, and the Internal Revenue Service.

There are many financial ratios used by contractor stakeholders, including equity partners and the bank. These ratios compare entries from both the balance sheet and the income statement. These mathematical ratios are a gauge of the contractor's financial health at several levels. Various financial management reports are generated by the contractor's home office accounting department that assist the jobsite team with cost control and forecasting. If there are errors in these reports, they must be journaled out and moved to correct cost codes as quickly as possible to allow the jobsite team to see a true picture of its cost status. Journal entries are not an actual movement of cash, but rather a paper debit to one code and credit to another code. The sum total of journal entries should balance out to zero, similar to the balance sheet.

The home office accounting department also produces financial reports which can assist the jobsite teams' cost control efforts including the job cost history report and weekly labor reports. There are numerous other ledgers and statements and schedules and worksheets that are part of construction financial accounting, but this discussion was limited to those which have the most effect on the jobsite team. In the next chapter on cost control, many of these financial statements are utilized by the jobsite team to endeavor to achieve their portion of the company's financial goals.

Review questions

1. Why do PMs, PEs, superintendents, and jobsite cost accountants need to bother studying financial statements and financial ratios which are generated by the home office accounting department and the CFO?
2. Owner's equity is a positive amount of money, at least the owners would hope so, yet it is regarded as a liability on the balance sheet. Why is that?
3. Why is the ROE ratio important to the officers of the construction company?
4. Starting with 'Revenue,' draft the long version of the income statement which includes all of the sub equations included in this chapter, and ending with 'Net profit.'
5. Of all of the financial statements discussed in this chapter (balance sheet, income statement, job cost history report, labor report, and equipment ledger) which one would the construction company owners be most excited about an increase in value?
6. Which of the five projects completed by ECC in 2020 as reflected in Table 7.2:

   a. Earned the highest total fee?

   b. Earned the highest fee percentage?

   c. Had the most effective use of jobsite general conditions?

7. Combining Tables 7.1 and 7.2, what was ECC's ROE and does that fall within the acceptable range?

8. Which ratio should be higher for a contractor which has been in business for over ten years, ROE or ROA, and why that choice?

9. The estimate for our hotel case study project presented in Chapter 4 was $24.5 million, but in Table 7.2 it is only shown as $9.8 million. Why the difference?

Exercises

1. Why do you suppose bonding and banking companies want contractors to increase their total assets and increase owner's equity?

2. Why would a contractor make a decision to own equipment either a) internally, b) through a separate equipment division, or conversely c) not own but rather rent from external sources? You may need to look ahead to Chapter 12.

3. This very successful general contractor always had revenue greater than $200 million, and a fair profit margin, yet the officers of the company endeavored to keep their total equity valued approximately $4 million: a) Where do you think all of the profits went? b) Why do you think they did not want OE to substantially increase?

4. How would you resolve the mis-management of jobsite financial affairs as discussed in Examples One, Two, and/or Three?

5. Provide two additional examples of each: a) short-term asset, b) long-term asset, c) current liability, and d) long-term liability.

8

Cost control

Introduction

As discussed earlier, construction is different from other industries. One way it is unique is that a contractor will not really know what a building will cost until it is 100% complete. The price is agreed to up front and the contractor contracts to that price, but has no guarantee they can build it for the amount agreed upon. This is one of the major reasons of having cost control. Some of the other reasons or purposes of cost control include:

- Requirement for an open-book construction contract;
- Preparing accounting statements including tax preparation;
- Monitoring employee performance;
- Tracking company assets such as construction equipment;
- Producing a profit for the company;
- Gauging what type of work the contractor does best and should specialize in; and
- Personal satisfaction.

This chapter discusses the cost cycle, beginning with development of the estimate and completing with input to the estimating database. The focus is again on the jobsite cost control efforts more so than the home office, but always relating what the project management team does at the jobsite to home office requirements. Buyout of subcontractors is included as a critical element of the second phase of cost control. The amount the general contractor (GC) contracts with subcontractors is not always exactly what the GC bids to the client or what the subcontractors bid to the GC. Work packages are a way to communicate the estimate and schedule and gather the tools and materials necessary to allow the foremen and craftsmen to efficiently build an element of the project. During construction, all expenditures for labor, materials, equipment rental, and subcontractor invoices need to be cost coded properly; otherwise the jobsite team will not really have a true picture of the cost of the work compared to how it was estimated. The project manager

(PM) and jobsite cost accountant input to the home office for all construction management (CM) control functions, including cost control, with a variety of reports. The monthly fee forecast is a tool that the chief financial officer (CFO) and chief executive officer (CEO) use to update corporate financial statements and communicate with external stakeholders such as bonding and banking strategic partners.

Cost control cycle and connection with accounting processes

There are five phases of cost control in a typical contractor's cost control cycle. These include *setup* or estimate preparation, *correct* or adjust the estimate after buyout and input to the cost control and accounting system, *record* or monitor costs including cost coding, *modifying* the system if costs are not being achieved, and preparation of an *as-built* estimate and input back into the company database for use in the next estimate.

The first phase begins with an accurate estimate and schedule. Only if the contractor is successful in its bid or proposal to the client will it move beyond phase one. If notified by the client of success, the first thing the contractor must do is validate the accuracy of the estimate, correct it as necessary through a variety of methods, and then enter it into the corporate cost control system. The home office accounting department will assign a unique job number once the contract has been signed. These first two phases of cost control are discussed in the next section of this chapter. A useful cost planning tool is the development of foreman work packages. During the course of construction, actual costs are recorded during phase three and entered into the accounting system,

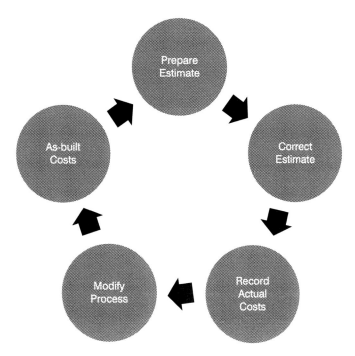

Figure 8.1 Cost control cycle

many as accounts payable. Phase four is a combination of discovering if there are problems with the project cost control systems, making adjustments as necessary, and reporting progress to the home office. Change orders must also be incorporated into the project plan and cost control and accounting systems. Phases three and four are also discussed in this chapter. The fifth and last phase of cost control involves development of the as-built estimate, which will be discussed in Chapter 16, 'Financial close-out of the construction project.' All five of the cost control phases are depicted in the cost control cycle in Figure 8.1. This same cost control figure will be discussed throughout this book on cost accounting and financial management.

Establishment of the cost base, including buyout

As discussed in detail in Chapter 4, the *estimate* is an assembly of measured material quantities, competitive market-rate material unit prices, historical direct labor productivity rates, current direct labor wage rages, competitive subcontract quotes, and a series of markups and fees. In order to better prepare the jobsite management team for cost control, the original estimate should be assembled by work packages and every line item in the estimate assigned an individual cost code.

No estimate or schedule is ever perfect. After the contractor receives notice of award of its bid or proposal, the estimate must be *corrected* with actual subcontract and purchase order buyout values and input into the company's cost control system. If the estimator made any mistakes, these must be corrected now, with potential journal entry modifications to contingency funds or fees. The superintendent and jobsite team cannot begin construction and affective cost control efforts with an incorrect estimate.

A major element of the estimate correction process, before it is input to the cost control system, is the buyout of subcontractors and suppliers. The buyout process begins first with detailed requests for proposals and/or quotations (RFPs and RFQs) issued from the general contractor to potential subcontractors. Subcontractors should be 'vetted' in that only qualified firms should be invited to bid. Once bids are received, then they should be compared beyond just the bottom-line price. Subcontractors should be invited to the GC's office, or the GC should visit the subcontractor's place of business, and the drawings and specifications should be reviewed in detail. Any potential bid qualifications or exclusions should be agreed upon now, rather than debate over potential gaps in scope months from now when the project is under construction. Subcontract agreements and purchase orders should only be awarded to 'best-value' companies, those with complete prices, scopes, and quality and safety and schedule plans that meet the GC's expectations.

Foreman work packages

The major categories of any estimate include:

- Direct labor,
- Direct material,
- Subcontractors and major material suppliers,
- Jobsite administration or general conditions, and
- Percentage markups including fee, excise tax, contingency, and insurance.

The markups, including home office overhead, are beyond the control of the jobsite team and are the focus of the CEO and CFO. If subcontractors and suppliers were bought out diligently, and tight agreements written, then there is little to 'control' with respect to their costs. Direct materials should have been accurately measured and competitive prices applied to those quantities. The superintendent cannot install less concrete in the footings to save money than was drawn, so cost exposure is limited for materials as well. Most of the items within the jobsite general conditions estimate can be managed, but the biggest variable there is time – if the project lasts longer, jobsite general conditions will run over. Construction equipment should be cost-coded to direct work and not general conditions wherever possible. Efficient management of equipment is necessary and is discussed in Chapter 12, 'Equipment use and depreciation.' *Direct labor is the most difficult variable* for the estimator to forecast and is the most difficult for the superintendent to 'control' and therefore deserves the majority of focus of any discussion on construction cost control.

Control or management of direct craft labor and equipment rental cost is the responsibility of the superintendent, often with the assistance of a field accountant or jobsite cost engineer. The key to getting the foremen and superintendents involved in cost control is to get their personal commitment to the process. One successful way for the general contractor to do this is to have the superintendent actively involved in developing the original estimate. If the superintendent said it will take five carpenters working three days to form the spot footings, he or she will endeavor to see to it that the task is completed within that time.

Some construction executives believe that superintendents should not be told the true budgeted value of each work package. This is a poor practice, because the foremen and superintendents are key members of the project team and have critical roles in achieving project financial success. These men and women are regarded as the last planners and should be provided the actual budgeted cost for installation, both in materials and man-hours (MHs), as well as the scheduled time for installation for each work package.

One very broad technique for monitoring project cost is to develop a whole project direct work labor curve similar to the one shown in Figure 8.2 for the Olympic Hotel case study. For

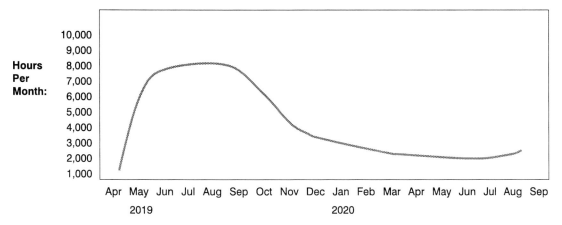

Figure 8.2 Whole project direct work labor curve

this example, all of the direct work has been combined into one curve; it is not separated by craft or work package as will be discussed later. It is important to have the superintendent or foremen record the actual hours incurred weekly and chart them against the estimate; the jobsite cost engineer can help with this. If the actual labor used is under the curve it is assumed the foremen and their crews are either beating the estimate or behind schedule. The opposite is true if the hours are above the curve. This simple method of recording the man-hours provides immediate and positive feedback to the project team, but it has its limitations as will be discussed in the next chapter on earned value. It is better to use hours and not dollars when monitoring direct labor. This is one advantage of estimating with MHs over unit prices for labor. Foremen and superintendents think in terms of crew size and duration. They do not think in terms of $100 per each to install bath mirrors, rather they have scheduled one carpenter to work three hours on the task.

Many in the construction industry, especially consultants such as estimators, think in terms of contract document organization according to strict Construction Specifications Institute (CSI) format. CSI division 03 is for concrete, and all cast-in-place (CIP) concrete is lumped together and not formatted in the estimate according to systems or assemblies. Therefore all concrete foundations, slabs-on-grade (SOGs), columns, elevated decks, and walls are grouped together. This may even include grouping off-site precast concrete with on-site CIP or tilt-up concrete. Conversely, any work which may be specified in another CSI division, such as structural excavation and backfill or footing drains or structural steel embeds, will not be grouped with the concrete work, even though these activities are performed with the CIP concrete.

Preparation of the original estimate, schedule, and supporting work breakdown structure (WBS) are more efficiently done by work packages or assembles than a pure CSI approach. Work packages are a method of breaking down the estimate into distinct packages or assemblies or systems that match measurable work activities. For example, concrete footings, including formwork, reinforcing steel, and concrete placement could be a work package. The work is planned according to the amount of hours in the estimate and monitored for feedback. When the footings are complete, the superintendent and the management team will have immediate cost control feedback.

This advantage of planning construction by assembly or work packages, in lieu of CSI, is evident in other areas of direct work as well. When an assembly such as wood framing is complete, which is part of CSI 06, the contractor will have immediate feedback as to how the original budget and schedule look. If all of CSI 06 was grouped together, including rough carpentry, blocking and backing, finish millwork, and cabinetry, the contractor would not necessarily know how the original budget was looking until the entire project was complete.

Which items should the project team track? The 80-20 rule applies here and elsewhere in cost accounting and financial management. With 80% of the cost and risk lying with 20% of the activities, it is those activities that deserve the most cost control attention. The jobsite project team should evaluate and mitigate the largest risks. The estimate should be reviewed to identify those items or systems that have the most direct work labor hours. Work packages should be prepared for those items the cost control team believe are worthy of tracking and monitoring. Each work package should be developed by the foreman who is responsible for accomplishing the work, maybe with the assistance from the cost engineer. Some examples of simple straight-forward assemblies that warrant work packages on the hotel case study include:

- Spot or continuous footings,
- Cast-in-place concrete elevator pit,

- Garage slabs-on-grade,
- Elevated garage and first-floor lobby concrete decks,
- Doors/frames/hardware,
- Toilet accessories, including mirrors,
- Kitchenette cabinets and appliances,
- Exercise room equipment, and others.

Contractors think of work in terms of 'assemblies' or 'systems.' The concept of work packages involves grouping all items of work in one assembly, regardless of specification division, supplier, or craft performing the installation. In this way the contractor knows what that assembly costs and where it fits into the project schedule. The means and methods of construction belong to the contractor and particularly to the superintendent. The superintendent also efficiently assembles all of the materials, labor, and equipment necessary to construct the work. A work package for installing the hotel guest room door frames, hanging the doors, and installation of door hardware and door signage is used here as an example. In this scenario, the superintendent is meeting with Joe, his carpenter foreman, to plan the work and hand over the work package for assembly. Some of their discussion and considerations for an affective operation would include:

1. A list of suppliers and their points of contact, including phone numbers and email addresses, is shared with the foreman.
2. Supplier delivery dates are noted.
3. Approved submittals for all materials should be available in the job trailer.
4. Hollow-metal hallway door frames are to be grouted and an assembly area for each floor needs to be set up. They do not want to move grouted frames vertically in the building. These will be heavy, and the crew will be educated on safety precautions.
5. Fire rated frames and door leafs must be factory labeled.
6. The GC will set up a locked and controlled door hardware room in the shelled first-floor restaurant. Only carpenter foreman Joe and Roger, Evergreen Construction Company's superintendent, will have access.
7. Joe is told they have estimated 432 hours and three weeks for this work package.
8. There will be a crew of three carpenters installing this work and one additional laborer helping grout the frames, stage materials, and cleanup.
9. There are not any door relites to worry about in the hotel rooms but there are in some of the first-floor common areas.
10. Cost codes and estimated hours and schedule milestones are shared with the foreman.
11. The foreman will charge his time 100% to the work and will split his efforts between logistics for his crew and installing doors with his tool bag on.
12. There will be one toolbox for the crew located on the floor they are working on and locked up each evening.
13. One carpenter will move off the floor one day before the crew is complete and begin layout on the next floor.
14. The team will hold off on door signage until one week prior to punch list to avoid damage. Cardboard room numbers will be used as temporary signs. Twenty hours will be set aside from this work package for signage.

Project: Olympic Hotel and Resort
System/Area: Doors, Frames, and Hardware

Foreman: Joe Wallace
Date Developed: 11/20/19

Work Days:	1	2	3	4	5	6	7	8	9	10	11	12	13	14	15	16	17	Totals:
Estimated Crew Size:	3	3	3	3	3	4	4	4	4	4	4	4	4	4	3	0	0	
Estimated Daily Hours:	24	24	24	24	24	32	32	32	32	32	32	32	32	32	24	0	0	432
Accumulated Hours:	24	48	72	96	120	152	184	216	248	280	312	344	376	408	432	432	432	432
Actual Crew Size:	2	2	2	3	3	3	4	4	4	5	5	5	5	2	2	1	1	
Actual Daily Hours	16	16	16	24	24	24	32	32	32	40	40	40	40	16	16	8	8	
Accumulated Hours:	16	32	48	72	96	120	152	184	216	256	296	336	376	392	408	416	424	424

Manpower Curve:
Hours:

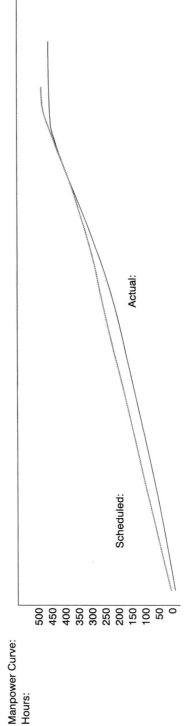

Figure 8.3 Foreman's cost control work package

There are likely additional details and information the foreman will need, but this is just an example of many of the elements that might go into one simple work package. Work packages for systems such as elevated concrete decks, structural steel, wood framing, siding, or roofing would be even more involved – but this process is all necessary to prepare a thorough work package to assure a successful construction operation. Figure 8.3 shows an example foreman's cost control work package. The detailed estimate for this system was included earlier.

Direct labor and material and subcontract cost coding

Cost control begins with assigning cost codes to the elements of work identified during the work breakdown phase of the cost estimate. Creation of the WBS is one of the first steps in estimate development. These cost codes allow the project manager, the cost engineer, and the superintendent to monitor actual costs and compare them to estimated costs. The objective is not that the team has to rigidly keep the cost of each element of work under its estimated value, but rather to ensure that the total cost of the completed project is under the estimated cost. Some uses of actual cost data include:

• Monitor project costs, identify any problem areas, and select mitigation measures;
• Identify additional costs incurred as a result of changes and process change order proposals;
• Identify costs for completing work that was the responsibility of a subcontractor and process associated back charges;
• Provide the project client with a cost report, which may be an open-book contract requirement;
• Evaluate the effectiveness of the jobsite project management team; and
• Develop a database of historical cost data that can be used in estimating the cost of future projects.

Cost codes

To be able to control costs, they must be tracked accurately and compared against the corrected estimate. The first step is to record the actual costs incurred and input the information into a cost control database. Cost codes are used to allow comparison of actual cost data with the estimated values. There are several types of cost codes used in the industry. The key to choosing a cost coding system is first to make sure that all projects within the company use the same codes, so that actual costs may be compared upon project completion and input into a usable estimating database. The cost code system should also be the same as other codes including file codes, subcontract and purchase order codes, and estimating and scheduling codes. An example of a cost coding system would include elements of the specific job number, a designator of labor versus material versus subcontractor cost, and the specification number. Some contractors will use RS Means cost codes, which are similar to Construction Specifications Institute codes.

Depending upon the size of the construction firm, the type of work, and the type of client and contract agreement, the general contractor may perform job cost accounting in either the home office or in the field. Generally, the smaller the firm and the smaller the contract value, the more likely all accounting functions will be performed in the home office. On larger projects, the project team may have a jobsite cost accountant. The type of contract and how it addresses reimbursable costs may also have some effect on where the construction firm performs cost accounting. Two different examples follow.

Example One: This project is a $100 million downtown office building that has a negotiated guaranteed maximum price contract that allows for all on-site accounting to be reimbursed. The project is only four blocks from the GC's home office and it may be more cost effective to perform accounting out of the home office with the assistance of an accounting department, but according to the terms of the contract, the client will not pay for activities conducted off the project site, therefore the accountant is in the jobsite trailer.

Example Two: This is a $50 million competitively bid lump sum school project located ten miles out of town. Since this project is closed-book, any costs not expended on the project site transfer into increased fees and ultimately increased profits for the construction company. Therefore any accounting activity that can be accomplished more effectively in the home office is done so there.

Regardless of where the cost data is collected and where the checks are prepared, most of the accounting functions on a project are the same. The process begins with a corrected estimate. Then actual costs are incurred, either in the form of direct labor, material purchase, or subcontract invoice. Cost codes (those matching the estimate) are recorded on the time sheets and invoices. Often this process begins with the cost engineer. The coded time sheets and invoices are then passed to the superintendent and the project manager for approval. Sometimes the officer-in-charge or maybe the client (on cost-plus projects) may also want to initial approval on each invoice. After the time sheets and invoices are coded and approved, the cost data is input into the cost control system.

One important aspect of this phase of cost recording is the accurate coding of actual costs. If costs are accidentally or intentionally coded incorrectly, the project team will not really know how they are doing on that specific item of work. Some superintendents may have their cost engineers intentionally code costs against items where there is money remaining, not necessarily against the correct work activity, thereby hiding over-runs. The jobsite staff will not be able to monitor and correct construction processes if coding errors occur. All costs should be coded correctly to provide the project team an accurate accounting of expenditures. These next two examples reinforce the need for accurate cost coding.

Example Three: This cost engineer was required to review time sheets daily for 1,000 pipe fitters on a very large power plant project. One particular pipe hanger seemed to have 300 or so man-hours charged to it daily. On further inspection by the cost engineer, that hanger was discovered right above the door exiting the turbine building. The pipe and hanger had been there for years and had cobwebs on them. But what was very clear for all of the fore-men looking for an easy cost code as they finished their work day and exited the building was this one pipe hanger number.

Example Four: This medical facility suffered a devastating fire when it was nearly 70% complete. Creation of the cost estimate for recovery was very complicated. The estimator had insurance agents and the client and his bosses and all of their attorneys looking over his shoulder during estimate preparation. All parties were concerned about mixing the hard-cost, which had not been expended for the original contract, with the costs for remediation. A separate jobsite cost accountant was brought in once the estimate was finished to cost code only those costs for repair. The accountant's wages were added to the insurance claim, which increased that price, but all parties felt more comfortable that the accounting books would be kept separate.

Cost recording is the largest and most time-consuming phase in the cost control process. Recording actual costs is a function most-often performed by project engineers and jobsite cost accountants if available. Some of the ways the cost engineer is involved in recording costs include:

- Assist foremen and superintendents with development of work packages;
- Enter direct labor cost codes on time sheets;
- Enter cost codes on short- and long-form purchase orders and subcontract agreements;
- Verify that supplier and subcontractor monthly invoice amounts match their contract values;
- Enter cost codes on supplier and subcontractor invoices; and
- Assist the project manager with monthly fee forecast development.

Monthly project management forecasting to the home office

The degree of autonomy any jobsite project team has from the home office varies depending upon the culture of the construction company, the complexity and size of the project, contract terms, and the individual team members involved. In the hotel case study, the distance the jobsite was from the home office also played a major role, especially with respect to jobsite cost accounting. Regardless, there will always be input and guidance from the home office to the jobsite as well as reporting accountability from the jobsite back to the CEO and CFO.

As stated, there is rarely a 'perfect estimate'; many construction projects will not proceed 'exactly' according to the original plan and schedule, despite management's good intentions. Continually comparing actual costs recorded against the estimate will discover *variances which warrant attention* and potential adjustment by the project team. The contractor cannot wait until the project is finished to find out if it made money. At that time there is nothing that can be done to fix the problem. Costs may exceed the estimate for a variety of reasons, including:

- The estimate was in error,
- Cost coding or data entry errors,
- The project team is not implementing construction management processes according to the original plan,
- Subcontractors or suppliers may not be performing,
- Direct craft field foremen may not be performing,

- Project management team members, including the PM and superintendent, may not be performing,
- The general contractor is installing work which was the responsibility of a subcontractor or the client,
- The client or designer may have added work,
- Inclement weather,
- Unforeseen conditions are discovered,
- The project is behind schedule, or
- Excessive overtime used to maintain the schedule.

Once the reason for the cost over-run is understood, the plan or process may need to be *adjusted or modified* by the jobsite team and hopefully the cause was discovered soon enough to implement an adjustment. Some of the corrective actions, or 'fixes,' might include:

- Utilize different means and methods or equipment,
- Enforce subcontract agreements,
- Back charging subcontractors,
- Change in-house general contractor personnel,
- Change subcontractors (this is difficult to do),
- Change order the client,
- Expedite the schedule, or
- Selective use of overtime – sometimes spending money can save money.

The project manager is responsible for developing a *monthly forecast* for the project that will be shared with the superintendent and the officer-in-charge. This should be prepared with the assistance of the project cost engineer and field foremen and superintendents. The contractor's bonding and banking stakeholders may also have an interest in the monthly forecast. The client may be copied in with the monthly forecast in the case of a negotiated cost-plus contract. This forecast includes line items for all areas of the estimate, cost to date, and estimated cost to complete. Each of the major areas of work receives a separate forecast page, and each of those broken down for all categories of work. As stated, the major categories of the estimate include direct labor, material, equipment, subcontractors, and jobsite general conditions.

The process in developing the cost forecast is a series of mathematical calculations and each one provides a more accurate projection of what the costs will be at construction completion compared to the original estimate before the project started. The forecast spreadsheet utilizes a simple series of Excel rows and columns and follows the format shown later in Table 8.2. Development of the monthly project management cost forecast includes the following steps:

1. Start with the original or contracted estimate. The sum total at the bottom of this column should exactly equal the original contract value.
2. The next column incorporates approved change orders only. Unapproved change orders or change orders which are in process should not be reflected here as there is no guarantee that they will be incorporated into the contract value.
3. The fifth column sums columns three and four and totals the current contract value.
4. The sixth column provides actual costs to date as reported from the monthly job cost history report.

5. The next column is the most difficult. All of the previous columns have been simple cut and paste or forwarding from other accounting reports and routine math calculations. The seventh column is the 'forecast' column where the project team must make an educated prediction of how much money is left to be spent. The optional ways to develop the forecast include:

   a. Simply inserting the amount of money contracted but not yet spent such that the individual line item will total out perfect and there will be a variance at the end of $0.00. This can be done by setting up the columns to automatically compute the to-go cost. For example, if the original estimate was $2,500, and $2,200 has been spent, then there is $300 to go.

   b. Using a trend analysis such that if the team has spent $500 per cubic yard (CY) of concrete on the first 1,000 CY, and they have 500 CY to go, then they have 500 CY at $500/CY or $25,000 left to spend. This can also simply be created using math formulas.

   c. Utilize the initial productivity rate times the quantity to go, assuming that there was some good thought put into that original rate. If 14 man-hours (MH) per 1,000 board feet (MBF) were estimated for 100 MBF of rough framing, but 16 MH/MBF have been incurred on 80 MBF thus far, then 14 MH/MBF, the original productivity rate, is multiplied times the to-go quantity of 20 MBF with no consideration given to the rate trended.

   d. The most accurate way to compute costs to go is to understand what the quantity is to go and make an educated assessment for the productivity needed to finish that activity with input from the foremen and superintendent. This should be done on a line by line basis.

6. The first three of these methods are easy to do, but they all are invalid to some degree. There may have been learning curve issues. There may be cost code errors. There may be changed conditions. Some forecasting methods may be used on some line items and others in other locations. Examples of all of these forecasting options, for structural steel beam installation for the Olympic Hotel, including over- and under-running the estimate scenarios, are shown in Table 8.1.

Table 8.1 Forecasting options

Forecasting	Estimated			To-date			To-go			Total	Variance
Option	Qty	UMH	MH	Qty	UMH	MH	Qty	UMH	MH	MH	+/–
Steel beam over-run scenario:											
A	24	2	48	18	2.5	45	6	0.5	3	48	0
B	24	2	48	18	2.5	45	6	2.5	15	60	–12
C	24	2	48	18	2.5	45	6	2	12	57	–9
Steel beam under-run scenario:											
A	24	2	48	6	1	6	18	2.3	42	48	0
B	24	2	48	6	1	6	18	1	18	24	24
C	24	2	48	6	1	6	18	2	36	42	6

Notes:
Forecasting option 'A' forecasts the original total estimate.
Forecasting option 'B' forecasts the current productivity trend.
Forecasting option 'C' forecasts the original productivity rate.
A variance of - MHs indicates an over-run in cost and a + MH indicates an under-run in cost.

7. The eighth column is the total forecast which adds column six, cost to date, with column seven, cost to go. The total at the bottom of the total forecast column reflects what the project team estimates the project will cost when complete, given the updated information they now have. Each month this forecast becomes progressively more accurate.
8. The last column is a variance column. It subtracts the total forecast from the approved and revised contract values. A positive figure represents the contractor is under-running the estimate and will improve its fee position; a negative number (often red) indicates a projected cost over-run.
9. The final calculation in the forecast is to project the contractor's fee based upon the compilation of all of the individual line item variances from column nine. The home office tasks the project team to bring in the fee that they originally bid, if not improve upon it.

Two additional columns could have been added to the forecast worksheet between columns three and four to incorporate estimate correction journal entries and subcontractor buyout but were omitted from this example for simplicity as the total of journal entries would have been zero.

Table 8.2 illustrates the monthly forecast summary page for the hotel case study project. The complete forecast is often as long as the original detailed estimate; possibly 10–20 pages and hundreds of line items. It is a good practice to include a narrative with the monthly forecast explaining the significant differences from the previous month's forecast, as well as a work plan for continuing or improving performance throughout the remainder of the project. One important rule of construction cost accounting is that the financial reports must be consistent to allow the home office a clear picture of project progress as well as for them to develop confidence in the jobsite team. The management team cannot afford to wait until the project is complete to measure and report the overall project cost. It is not only too late to take corrective action, but it is also too late to accurately determine why the team deviated from the plan. The following example is from an overly pessimistic project manager.

Example Five: This project manager had been a project engineer for about seven years, had managed a couple of small projects as a PM, and was now managing a high-risk lump sum bid $15 million aerospace project with an estimated 3% or $500,000 fee. He was 'chicken-little' with a 'sky-is-falling' attitude throughout the whole project. His fee forecasts varied each month from $0 to $200,000 in the black, to $200,000 in the red. The result was the home office had a senior PM, and an additional staff superintendent, visit the site once weekly for the final three months. He had been overly pessimistic with his forecasts and thought himself quite the hero when the project settled on a $1 million fee. He was terminated shortly after final close-out.

There can be several other management reports either generated by the jobsite team or the home office accounting department according to the construction firm's practices and the requirements of any specific client or project. Typical cost accounting and financial management reports include the weekly labor report, monthly job cost history report, equipment log or ledger, and accounts payable report. Most of the reports are computer generated and are accurate to the degree that the information regarding actual costs was accurately input. They can occur weekly or monthly, but,

Table 8.2 Monthly forecast summary

Evergreen Construction Company
MONTHLY FORECAST SUMMARY PAGE

Project: Olympic Hotel		Job No.:	422	PM:	Chris Anderson		Date:	1/1/2020
Col. 1	2	3	4	5	6	7	8	9
CSI Division	Description	Estimate Totals	Change Orders (2)	Current Contract	Cost To Date	Cost To Go	Forecast Cost	Variance +/–
1	General Conditions	$1,662,783		$1,662,783	$845,450	$831,000	$1,676,450	–$13,667
3	Concrete	$4,826,612	$68,853	$4,895,465	$4,545,450	$180,005	$4,725,455	$170,010
4	Masonry	$150,000		$150,000	$0	$150,000	$150,000	$0
5	Structural Steel	$411,103		$411,103	$405,045	$2,500	$407,545	$3,558
6	Carpentry	$3,119,193		$3,119,193	$2,316,000	$896,000	$3,212,000	–$92,807
7	Waterproofing and Roof	$1,394,814	$20,000	$1,414,814	$315,000	$1,089,814	$1,404,814	$10,000
8	Doors and Windows	$1,284,940		$1,284,940	$425,000	$859,940	$1,284,940	$0
9	Finishes	$2,337,066	$50,000	$2,387,066	$0	$2,387,066	$2,387,066	$0
10	Specialties	$157,085		$157,085	$0	$157,085	$157,085	$0
11	Equipment	$258,550		$258,550	$0	$258,550	$258,550	$0
12	Furnishings	$96,687		$96,687	$0	$96,687	$96,687	$0
13	Special Construction	$0		$0	$0	$0	$0	$0
14	Elevators	$435,000		$435,000	$125,000	$330,000	$455,000	–$20,000
21	Fire Protection	$442,700		$442,700	$90,000	$352,700	$442,700	$0
22	Plumbing	$1,211,379		$1,211,379	$375,000	$826,990	$1,201,990	$9,389
23	HVAC	$720,629		$720,629	$115,000	$610,050	$725,050	–$4,421
26	Electrical	$1,286,702		$1,286,702	$725,000	$599,000	$1,324,000	–$37,298
27	Low Voltage	$332,500		$332,500	$0	$332,500	$332,500	$0
31	Site Work	$1,428,311	$37,564	$1,465,875	$650,000	$750,000	$1,400,000	$65,875
	Total Cost:	$21,556,054	$176,417	$21,732,471	$10,931,945	$10,709,887	$21,641,832	$90,639
	Labor Burden:	$1,074,287	$0	$1,074,287	$750,000	$315,000	$1,065,000	$9,287
	Fee @ 5%	1,150,833	$8,821	$1,159,654	$584,097	$551,244	$1,135,342	$0
	Other Markups	$748,841	$7,057	$755,898	$380,247	$358,860	$739,107	$0
	Total Contract:	$24,530,015	$192,295	$24,722,310	$12,646,290	$11,934,991	$24,581,281	**$99,926**
	Forecast Loss:			$0				
	Forecast Savings:			$99,926				
	GMP Savings Split: 80% Client, 20% GC =			$19,985				
	Current Contract Fee:			$1,159,654				
	Forecast Final Fee:			**$1,179,639**				

either way, home office generated cost control reports are likely to be too late for implementing any corrective action in the field.

Summary

There are five steps to the construction cost cycle, beginning with estimate setup, adjusting or correcting including buyout, cost recording once construction starts, modifying the system if needed, and finishing with preparation of the as-built estimate once the project is complete. Work packages are a way for site supervision to plan their work and implement cost control. In order to prepare these the foremen must first be given the cost estimate in terms of direct work hours and time to complete. Work package implementation also requires access to proper construction tools and materials for installation.

Cost codes follow the cost cycle from start to completion. The same cost codes should be shown on the estimate pricing recapitulation sheets, added to subcontracts when executed, noted on time sheets and material invoices, and input back into the corporate database when the as-built estimate is complete. If the jobsite financial team deems that they are missing their cost goals or schedule for any reason, they need to analyze the cause, and affect immediate remedy. This may require modification of the process, different equipment, or maybe a change of personnel is necessary. To wait until financial reports generated by the home office reach the field is too late. This is another advantage to utilize work packages and perform as much of the cost accounting as is feasible at the jobsite.

The project manager produces a monthly cost and fee forecast report for the home office. This is essentially a complete new estimate for the project, which incorporates costs to date with forecasted costs to complete. Each line item must be analyzed independently and factor issues such as quantity to go and learning curves. It is very likely that the CEO and CFO will share project forecasts, at least on a summary level, with internal equity partners and external stakeholders such as the bank and the surety. Consistency and accuracy are therefore paramount with construction cost reporting. Earned value is another cost control tool that is usually coupled with the material included here, but the next chapter is entirely devoted to that more advanced topic.

Review questions

1. Why should the PM 'correct' the estimate?
2. Define 'best-value' subcontractors.
3. Next to direct labor, what area of the estimate is difficult to estimate and therefore risky?
4. Looking back to the definition of a GC versus a CM in Chapter 3, what does the CM do to minimize the risk of direct cost over-runs?
5. What is an advantage of estimating with man-hours over unit prices for direct labor?
6. Explain the 80-20 rule as it relates to cost control.
7. When is forecasting option 5.a appropriate for a single line item in the budget?
8. When is forecasting option 5.b appropriate for a single line item in the budget?
9. Looking at the actual versus planned door, frame, and hardware work package in Figure 8.3, how did this system turn out with respect to cost and schedule?

10. How might inaccurately cost-coded time sheets affect a construction firm's ability to estimate accurately?
11. What is the difference between 'manage' and 'control' with respect to cost control?
12. How frequently should a PM develop a cost and fee forecast report?

Exercises

1. Other than the items bulleted earlier, list another reason costs may be exceeding the estimate and an alternate solution for that problem.
2. What might be missing from the door, frame, and hardware work package assembly discussed earlier?
3. Prepare a work package list of items and considerations for either a) exercise room equipment installation, or b) site work signage and furnishings including fountains.
4. What would happen if your superintendent discovered you had given him or her an erroneously low budget or tight time frame and you as the PM knew you had more to work with?
5. Planning construction by work packages is not limited to GCs. Using the hotel case study as an example, break down the work for either the painting subcontractor, gypsum wall board subcontractor, or roofing subcontractor into multiple distinct work packages or assemblies or systems. Make whatever assumptions are necessary.
6. Provide an explanation of how and why the door, frame, and hardware work package in Figure 8.3 might have followed the course it did and turned out in that fashion.
7. Referring back to the forecasting options from Table 8.1, which is the most correct, Options A, B, or C, with respect to a) over-running the estimate, and b) under-running the estimate? There is not an exact answer here, but some are more correct than others.
8. Look ahead to the abbreviated finishes database in Table 16.1. Instead of four different jobs, assume that this was all on one project but the data reflects four different foremen. Prepare scenarios for why a foreman may be excessively higher or lower than the others and what should be done with this data if discovered during the middle of a project.
9. A) Utilizing another case study project, prepare a work package for direct labor for a system of work such as concrete foundations, CIP concrete walls, tilt-up concrete walls, or SOG. Attach the original portion of the estimate that refers to this work package. B) As an alternate to a separate case study project, the following quantity and man-hour columns can be used as a baseline estimate. Assume a three-person crew working for four months from start to completion for all of these activities:

	Estimated quantities	Estimated hours
Excavation:	100 CY	100
Formwork:	5,500 SF	550
Reinforcement steel or mesh:	30 Tons	300
Concrete placement:	400 CY	200
Slab finish:	35,000 SF	350
2 × 6 wall framing:	33 MBF	500

10. Prepare a cost forecast for the work activities from Exercise 9. Forecast the remaining hours needed and calculate the total over- or under-run this project will achieve, assuming the following percentages complete and spent. Convert the hours to dollars using wage rates from Chapter 18 or other local current rates. Use whatever craft mix you feel is appropriate. List out what possible reasons the over- or under-runs could be occurring and what corrective actions should be taken. If all continues on this same trend (proceeds at the same rate as it has been), calculate what the historical as-built unit man-hours will be.

	Portion of work complete	*Portion of estimate spent*
Excavation:	100%	90%
Formwork:	90%	100%
Reinforcement steel or mesh:	80%	70%
Concrete placement:	50%	35%
Slab finish:	75%	85%
2 × 6 wall framing:	0%	5%

9

Earned value analysis

Introduction

Earned value (EV), also known as earned value management or earned value analysis, is one technique used by some contractors to determine the estimated value of the work completed to date (or earned value) and compare it with the actual cost of the work completed. The most effective use of EV to a general contractor's (GC's) jobsite financial management team is to track direct labor, which represents the construction contractor's greatest cost risk. Man-hours (MHs) and not labor cost is best used for monitoring direct labor as every project and every crew will have a different mix of apprentices, journeymen, and foremen and associated wage rates.

This chapter will discuss origination of the third curve, the earned value curve. Previous to this there was only the scheduled or estimated curve and the actual cost curve. The EV curve and process is an advanced measure of cost and schedule control for contractors. There are many formulas and indices associated with a study of EV, and a few of these will be introduced here as well.

In addition to cost control, earned value can also be utilized with pay requests. Some sophisticated project owners may choose to pay their contractors on an EV system rather than an estimated or actual expenditure-based system. Many textbooks on construction management and cost accounting combine discussions of EV with cost control, but an expanded EV discussion was warranted in a book devoted to construction cost accounting and financial management.

Development of the third curve

The work package curve developed in Chapter 8 charted the man-hours planned by the foreman and the actual MHs used for the Olympic Hotel and Resort case study door, frame, and hardware installation. During the first week when the actual MH curve was below the estimated curve, did this necessarily represent that the foreman was under budget? Could he have been behind schedule? The project team could actually be ahead or behind schedule and over or under budget because the work package compared only estimated to actual hours and did not consider the amount of work accomplished.

The addition of a third, or earned value, curve combines forecasted with actual cost and schedule performance which allows true productivity measurement. The EV curves plots the quantity of work performed by the foreman and his or her crew against the hours which were originally estimated to install that quantity. This third curve is determined by plotting the total number of man-hours estimated for the work package multiplied by the cumulative percent completed. If 32% of the hotel room toilet accessories have been installed by day four, then 32% of the 325 estimated hours have been 'earned.' In other words, 104 hours of the 325 hours estimated have been earned and 104 is plotted on day four. When the activity is complete, 100% of the 325 hours will have been earned – 325 hours was originally estimated and the contractor cannot earn any more or less than this. If 300 hours were actually spent at the completion of this work package, they will have realized a profit and conversely if they spent 400 hours, they will have lost money on this activity.

In a pure material quantity perspective, if there are 200 windows in the hotel, and 150 have been installed, then 150/200 or 75% of the total window estimate has been earned, regardless of how much was spent on those 150 windows. With this third curve, an actual measurement of productivity can be made. This method of monitoring will provide more accurate feedback to the project team for appropriate cost and schedule reporting and correction if required.

Earned value as a cost control tool

Two different metrics of measurement are now available using all three curves in concert; they are schedule and cost performance. The questions the third earned value curve answers are:

- Are we right on schedule, ahead of schedule, or behind schedule and by how many days?
- Are we right on budget, over budget, or under budget and by how many hours?

There are nine different combinations of potential answers to these two questions as reflected in Table 9.1.

Schedule status

The schedule status is determined by subtracting the actual time used to perform the work from the time scheduled for the work that has been performed. This is the same as measuring the horizontal distance between the earned value and the estimated curves shown in Figure 9.1. This work package

Table 9.1 Earned value performance matrix

Schedule Status:	Ahead of Schedule	On Schedule	Behind Schedule
Cost Status: Under Budget	1	2	3
On Budget	4	5	6
Over Budget	7	8	9

Work Days:	1	2	3	4	5	6	7	8	9	10	11	12	Totals
Estimated Hours/Day	16	16	24	24	30	30	30	30	30	32	32	31	
Cumulative Hours	16	32	56	80	110	140	170	200	230	262	294	325	325
Actual Hours/Day	16	24	32	32	32	34	40	40					
Cumulative Hours	16	40	72	104	136	170	210	250					250
Earned Hours/Day	16	24	30	32	32	32	32	32					
Cumulative Hours	16	40	70	102	134	166	198	230					230

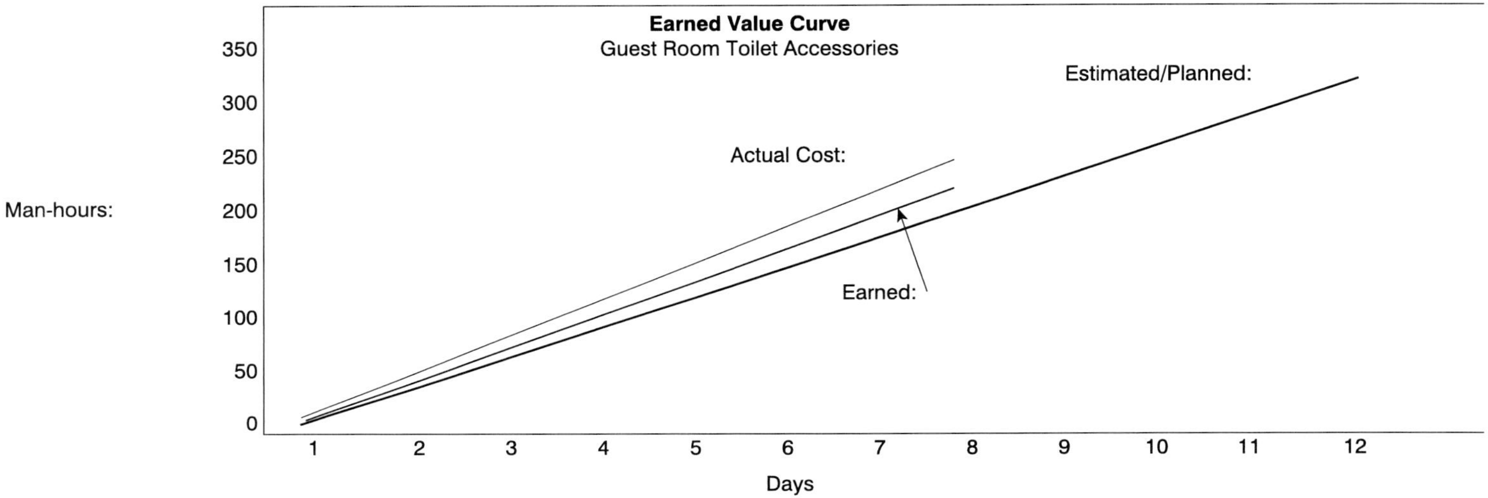

Earned Value Curve
Guest Room Toilet Accessories

Estimated/Planned:

Actual Cost:

Earned:

Days

Man-hours:

Simple Analysis:

Cost: On day eight, the foreman had spent 250 hours, but earned 230, therefore he was 20 hours over budget. This is the vertical measurement between the actual and earned curves.

Schedule: On day eight, the foreman had earned 230 hours but was not scheduled to have done so until approximately day nine, therefore he is approximately one day ahead of schedule. This is the horizontal measurement.

Figure 9.1 Earned value work package

Work Days:	1	2	3	4	5	6	7	8	9	10	11	12	Totals
Estimated Hours/Day	32	32	32	32	32	34	34	34	32	32	32	32	390
Cumulative Hours	32	64	96	128	160	194	228	262	294	326	358	390	
Actual Hours/Day	22	30	30	30	30	30	30	30					232
Cumulative Hours	22	52	82	112	142	172	202	232					
Earned Hours/Day	32	40	40	0	0	0	60	60					232
Cumulative Hours	32	72	112	112	112	112	172	**232**					232

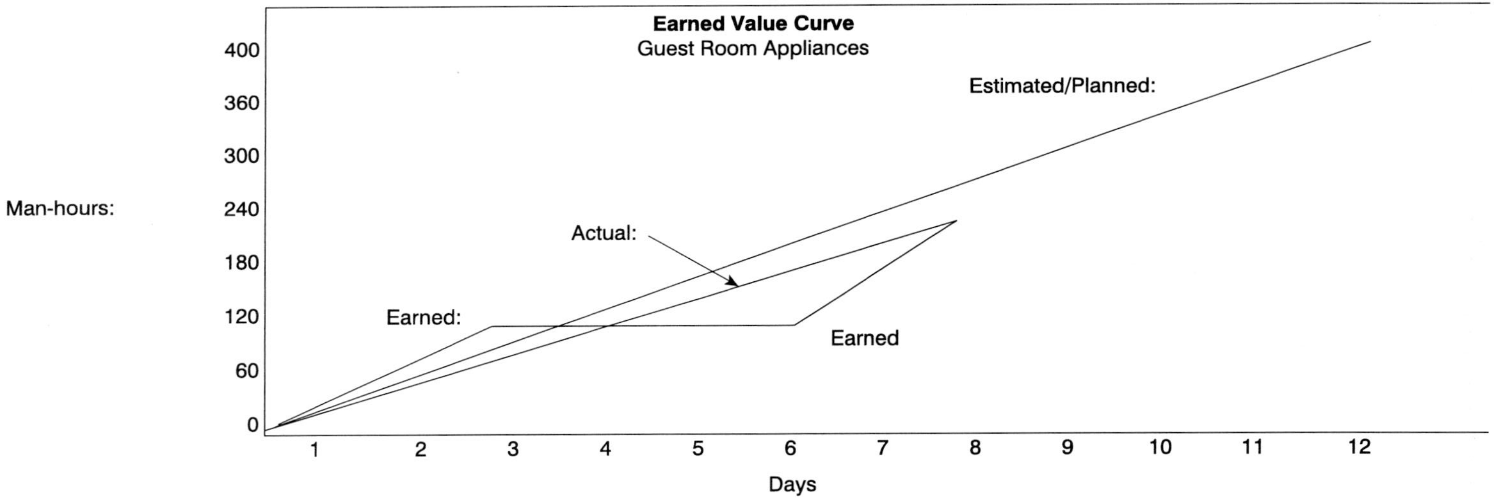

Earned Value Curve
Guest Room Appliances

Simple Analysis:

Cost: On day eight, the foreman had spent 232 hours, and also earned 232, therefore he was right on budget. This is the vertical measurement between the actual and earned curves.

Schedule: On day eight, the foreman had earned 232 hours but was scheduled to have done so approximately by day seven, therefore he is approximately one day behind of schedule. This is the horizontal measurement.

Figure 9.2 Earned value work package

is statused as of day eight. As of this point, the foreman had earned 230 hours of work, which was not scheduled to be in place until day nine. Therefore, the foreman is approximately one day ahead of schedule. This work package, turned EV analysis by the addition of the third curve, is for the hotel guest room bath accessories installation for the Olympic Hotel and Resort case study project.

Cost status

The cost status is determined by subtracting the actual cost of work performed from the 'earned' value of the work performed. This is the same as measuring the vertical distance between the actual and EV curves. Looking again at Figure 9.1, the actual cost of work performed was 250 man-hours while the earned value was 230 man-hours. Therefore, the foreman is 20 hours over budget. This places the foreman in option seven in the performance matrix Table 9.1 – ahead of schedule but over budget. Figure 9.2 for the microwave and refrigerator kitchen appliance work package shows a different situation. This foreman is on budget, but approximately one day behind schedule and is in option location six of the performance matrix Table 9.1.

Earned value indices

The previous curve comparison analysis is a straight-forward approach to reviewing the concept of earned value. Another way that many discuss EV is through a series of mathematical formulas and indices. In this case, earned and actual curves are not drawn, but points-in-time document the quantity of work installed and the cost of that work, either through man-hours as recommended for cost control, or through dollars as discussed later with pay requests. These points-in-time are represented in Figure 9.3. There are a multitude of acronyms and math formulas associated with a technical study of EV; here are just a few of them, along with a layman's explanation:

- ACWP: Actual cost of the work performed. This plots the actual point of hours or dollars when incurred, regardless of what was estimated.
- BCAC: Budgeted cost at completion. This reflects the total value of the work package at the end of the schedule, essentially the top point of the schedule or estimate curve.
- BCWP: Budgeted cost of the work performed. This plots the EV of work performed and when it was performed, regardless of actual cost.
- BCWS: Budgeted cost of the work scheduled. This is the estimated curve which reflects the original estimate in hours or dollars and when the work was originally scheduled to have been completed.

These acronyms all represent points that could be plotted on the graph. The EV indices compare these points utilizing the following mathematical calculations:

- CPI: Cost performance index = BCWP/ACWP, greater than one is under budget;
- CV: Cost variance = BCWP–ACWP, greater than zero is within budget;
- SPI: Schedule performance index = BCWP/BCWS, greater than one is ahead of schedule;
- SV: Schedule variance = BCWP–BCWS, greater than zero is ahead of schedule.

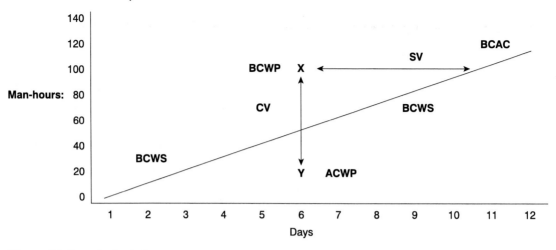

Figure 9.3 Earned value indices curve

Answering the questions of, "are we over or under budget and by how much and are we ahead or behind schedule and by how much?" can now be accomplished mathematically using these indices. The CPI for Figure 9.1 would be 230/250 = 0.92, which is less than one, and therefore over budget. Solving for CV is 230–250 = <20> which is less than zero and also over budget. The SPI for Figure 9.1 would be 230/200 = 1.15, greater than one, therefore ahead of schedule. SV also reflects ahead of schedule as 230–200 = +30 which is greater than zero. This confirms the approach that was used earlier, when we simply just compared horizontal and vertical differences between the three curves.

Material quantities can be used in lieu of hours or dollars with earned value analysis as well. The letter Q is inserted instead of C in the earlier acronyms, which reflects measurements of quantities scheduled and installed. These acronyms are presented in the following list. Quantity measurements may be based upon percent scheduled and completed. If there are 1,000 cubic yards (CY) of concrete estimated at an assembly unit price of $500/CY for a total of $500,000 and 300 yards of concrete are in place, then 30% (300/1,000) of the estimated $500,000, or $150,000, has been earned. The alternative is to track exactly how much concrete had been placed. This is sometimes done with a concrete log or by counting concrete trucks and multiplying their load size, such as 40 trucks times 9 CY per truck indicates 360 CY of concrete has been delivered and placed. Due to waste and spillage, this quantity of concrete may be slightly greater than the quantity reflected in the formwork.

- AQWP: Actual quantity of the work performed,
- BQAC: Budgeted quantity of the work at completion,
- BQWP: Budgeted quantity of the work performed, and
- BQWS: Budgeted quantity of the work scheduled.

Earned value as a pay request tool

The concept of payments made on an earned value basis was originated by the United States Department of Defense in the 1960s and adopted to construction in the late 1970s. To utilize EV in the pay request process requires a very sophisticated and construction-experienced client. This process can benefit a contractor which is on a stipulated sum contract and ahead of schedule or under budget. The lump sum contractor can also be hurt by this process if they are over budget or behind schedule. Guaranteed maximum price (GMP) projects with a pre-established schedule of values (SOV) operate very similar to a lump sum project if paid on an EV basis. This is assuming, though, that they are not subject to monthly audits of actual expenditures and corresponding adjustments to invoices. See Chapter 18 for additional discussion on audits.

Contractors with an open-book contract which are paid on a cost plus fee or time and material basis are usually not subjected to earned value for pay requests. If they were being paid on an EV process, and they were over-spending, they would be penalized; conversely they would be rewarded if they were under-spending as the owner would need to pay them in excess of what was spent and pay them what they had 'earned.'

If a contractor on a cost plus percentage fee is paid on a purely cost-plus basis and not earned value, and they are under-running their estimate, their fee would also be potentially reduced as fee in this case is based on a percentage of construction cost. Unfortunately this results in a disincentive for contractors to beat the estimate. Conversely if they were over budget, their fee would be increased in a pure cost-plus contract. This is why many cost-plus contracts include a high-side GMP protection for the project owner and 'fix' the fee at the time the estimate is developed. These contracts are also known as cost plus fixed fee.

Summary

The use of earned value analysis requires a thorough understanding of the interplay of work package planning with estimating and scheduling, along with accurate reporting of costs expended and quantities installed. With only two curves, the budget and the actual curves, whether they be man-hours or dollars, it is not possible to measure true performance on either cost or schedule. This would not be possible until project completion. The third curve allows the team to answer the interim questions: Are we on budget, under budget, or over budget, and by how much, and are we on schedule, behind schedule, or ahead of schedule, and by how much? The whole project direct labor curve introduced in Chapter 8 was a rough way to monitor cost, but the work packages developed by the foreman, which focus on a distinct assembly, usually organized by craft, such as carpenters versus electricians, is a much more accurate measurement. Now the third EV curve is even more accurate – but it requires more involvement from the jobsite cost engineer than the previous two.

The use of indices is a more technical EV analysis where formulas can be input into the computer and only percentage completion and cost or hours spent need to be added. Reporting only quantities is also an option. The use of indices without plotting the curve would also require additional input from jobsite cost engineers. Very few clients use EV as a method of payment, but actually invoicing lump sum projects with a pre-established schedule of values is very similar to EV. Using EV for pay requests is usually associated with experienced, sophisticated clients, such as the government for lump sum work, but not typically on privately financed negotiated cost plus fee projects.

Review questions

1. What is an EV analysis used for?
2. In which square in matrix Table 9.1, squares 1–9, would work package Figure 9.1 be at day four?
3. Figure 9.2 work package represents a more complicated scenario. What was the cost and schedule status as of days two, four, and six? Was the foreman ahead or behind schedule and by how much, and was he or she over or under budget, and by how much?
4. How is payment for a GC on an EV basis different than a lump sum basis with a schedule of values which was submitted on bid day and included in the contract as an exhibit?
5. If a GC was on a cost-plus project, without a GMP, how would payment on an EV basis affect them if they were a) under-running their estimate, or b) over-running their estimate?
6. If your project had estimated the foreman would spend 1,000 hours by June 1, but he or she had only spent 800 hours, is your team over budget or under budget or ahead of schedule or behind schedule? And by how much?
7. When would a contractor not want to use an EV curve for payment basis?
8. Which of the nine options in Figure 9.1 would be considered by a contractor to be a) most preferable, b) least preferable, or c) completely neutral?
9. What are the calculated CPI, CV, SPI, and SV for Figure 9.2 as of day eight and how does that compare to the results discussed in the narrative earlier?

Exercises

1. Prepare additional EV curves starting with the work package example presented from Chapter 8, Figure 8.3. Status the work package at day 12. Draw an EV curve where the foreman is over budget and behind schedule and another where the team is on budget and ahead of schedule. Prepare a narrative explaining why these situations might be occurring.
2. Referring back to Figures 9.1 and 9.2, if each of these work packages proceeds at the same trend as experienced over days seven and eight, when will they complete and what will be the final hours over or under budget?
3. In your opinion, what might have happened with the appliance work package in Figure 9.2? Why did the curve rise so quickly, flatten out, and then rise again?
4. In what option of Table 9.1 would the two points plotted in Figure 9.3 be? What if the ACWP and BCWP points were in the opposite locations?

10

Activity-based costing

Introduction

The topic of activity-based costing (ABC) has been a focus of many cost engineers, researchers, and academics since the early 1990s. ABC has been written on in several research articles and one recent book by Dr. Yong-Woo Kim, *Activity Based Costing for Contractors* (2017), and all cover the topic very well. This chapter just provides a brief history and relation to home office and construction jobsite applications for construction cost engineers and cost accountants.

Activity-based costing is not necessarily a process to reduce costs, but more one of identifying indirect costs so that they can be applied to the work allowing contractors to understand the true cost of direct construction activities. A study and application of ABC processes focuses on home office indirect costs, jobsite indirect costs, and fabrication facility costs. This will allow contractors to focus on improving costs where needed most and to specialize on those types of construction projects and construction activities which may be most profitable. Applying indirect costs to construction activities also allows a contractor to improve its estimating capabilities. It is generally understood that all contractors want to reduce costs, and the last two chapters on cost control and earned value have demonstrated some of the techniques that have been historically adopted. Controlling overhead costs has always been of interest to contractors, but the first step in ABC is to understand these costs and track them. In the next chapter, lean construction will be introduced which essentially takes the results from studying ABC and finds new and innovative methods where costs can be reduced.

Contractors typically allocate all home office overhead (HOOH) to their projects based on total annual revenue. If a contractor accomplishes $100 million of work annually, and incurs $2 million in HOOH expenditures, this amounts to 2% of their revenue. Each project will therefore also need to cover its share of HOOH. A $30 million project will need to cover 2% of $30 million, or $600,000 of HOOH. This does not factor whether every project requires an even share of home office support. This standard method of accounting is known as revenue-based costing. This method is appropriate for fixed overhead costs or organizational expenditures such as the chief executive's wages and the receptionist wages, as it would be difficult to attribute either of them to specific projects. Traditional accounting also does not consider economies of scale. For example, one $30 million project has one invoice per month requiring support from the home office accounts receivable clerk, the same effort he or she applies to a $1 million project for its monthly invoice. Indirect costs are not all shared equally and should not necessarily be totally proportioned based on revenue. The traditional revenue-based method also does not factor 'trouble' or risky projects which require additional attention from the company's general superintendent or quality or safety officers. Projects which are running smoothly and autonomously may require little attention from the home office.

Activity-based costing is loosely thought of as an approach to the costing and monitoring of activities which involves tracing resource consumption and costing final outputs. In other words, ABC applies costs incurred to what they produce. This chapter utilizes several examples to assist the reader to transfer from the traditional academic ABC study to practical construction company and construction project applications. Several examples are included of contractors which could benefit from a study of ABC, and others which have already been applying the concepts without necessarily recognizing it as so.

Foundation

Activity-based costing was initially popular with production industries, other than construction, such as the automobile industry, particularly by Toyota. ABC was studied in the mid-1990s for its potential application to construction. The construction industry and many of its participants have been historically 'set in their ways' and has not readily accepted a new approach to applying indirect costs such as proposed by ABC advocates. But a production industry application of ABC is easily applicable to manufacturing of construction materials such as structural steel, concrete reinforcement steel, and cabinetry, where production inefficiencies can be identified.

The philosophy behind activity-based costing is not necessarily how to reduce costs, but identify where costs should be applied. ABC tracks costs whereas lean construction is more how to reduce both overhead and direct costs. The answer is not necessarily to blindly eliminate or reduce overhead activities, as was mistakenly implemented in the next example, but more to understand what they are and how they are utilized.

Example One: On a competitively bid lump sum pharmaceutical manufacturing facility construction project, this electrical subcontractor was struggling to make its costs and due to damaged relations with the general contractor and the project owner, they were unable to receive all of their change orders approved, at least at full value. Towards the end of the project, when usually there is a spike in electrician manpower on any typical project, this company removed their project manager and superintendent to save costs and the foreman was stretched beyond his limits. The subcontractor declared bankruptcy soon after. Reducing general conditions activities sometimes costs more money than it saves.

It is worthwhile here to recall the revenue equation utilized throughout this book and its accounting derivatives, especially with respect to indirect costs.

$$
\begin{aligned}
\text{Revenue} &= \text{Cost} + \text{Profit} \\
\text{Fee} &= \textit{Home office overhead} + \text{Profit} \\
\text{Cost} &= \text{Construction cost} + \textit{Indirect cost} \\
\text{Indirect cost} &= \text{Home office overhead (HOOH)} + \textit{Jobsite general conditions} \\
\text{Construction cost} &= \text{Direct construction costs} + \text{Jobsite general conditions (GCs)} \\
\text{Direct cost} &= \text{Direct labor} + \text{Direct material} + \text{Equipment} + \text{Subcontractors}
\end{aligned}
$$

As a reminder, direct construction costs are those which are spent and traceable to actual construction activities, those which can be seen and touched on the jobsite. Direct construction costs include purchasing concrete and installing structural steel. Direct costs are easy to trace. Indirect costs are those which are either spread across company operations (in the case of home office) or spread across operations at the jobsite level. Examples of home office indirect costs would include the wages of the chief financial officer (CFO) and office building rent. Examples of jobsite indirect costs would include the project manager's (PM) wages and the trailer he or she offices in. Jobsite indirect costs are part of the cost of performing construction work and home office overhead is included with fee. It is also worthwhile to recall all of the different terms of general conditions which in some contexts are extremely different and others exactly the same:

• General conditions costs,
• AIA A201 contract document,
• Indirect costs,
• Home office GCs or overhead,
• Jobsite indirect or overhead or GCs or administration costs,
• General and administrative costs,
• Distributable costs, and
• Construction Specifications Institute (CSI) divisions 00 and 01.

The concept of activity-based costing has struggled with separating fixed from variable overhead and sometimes treats all fixed costs as variable. ABC applications should focus on home office variable costs. An acceptable method is to separate general conditions by those which are first 'organizational' – those that exist regardless of whether the company does $1 million or $100 million of revenue. The other category of GCs are 'assignable' costs which can be attributed to construction work, such as a concrete superintendent who assists project superintendents with planning their concrete pours and staffing needs.

It is the goal of some researchers and activity-based costing consultants to allocate all indirect costs, both home office and jobsite, to direct costs, but this is not feasible. Some overhead costs must be distributed over the company's operations, such as the wages of the CFO and other overhead costs which need to be distributed over all activities at the construction site, such as the superintendent's pickup truck. A realistic goal is to allocate as many variable costs to direct work as possible so the true cost of any element of work is known.

Methodology of activity-based costing

Activity-based cost accounting identifies 'value-added' overhead costs versus zero-value or low-value or resource-consuming overhead costs. ABC is similar to lean construction in this regard as lean focuses on eliminating these low-value resource-consuming costs. These differences may be identified through an ABC workshop customized for a specific contractor. Organizing and conducting an ABC workshop is very similar to a formal partnering session. Some participants may be worried that by tracking their activities their productivity or lack of productivity will be exposed, and therefore may be uncooperative with the ABC process. This needs to be addressed and resolved up front in the workshop. Some of the workshop's considerations include:

- Choose a facilitator,
- Budget the cost of the workshop,
- Develop a charter and mission statement,
- Stakeholders must attend and commitments must be made,
- Establish goals, and
- Identify indirect costs to study and cost drivers and individual construction activities, if down to that level.

Identification of *cost drivers* and construction *cost objects* is at the root of the ABC process. Cost drivers are those indirect activities which can ideally be allocated to projects or direct work packages. Cost drivers create or drive the cost and must be in measurable units. Cost drivers are in three categories:

- Transaction drivers: quantity of occurrences such as tracking copy machine use or long distance phone calls;
- Time drivers: durations applied to a project or task; or
- Budget drivers: cost or volume/revenue based.

The cost driver is the link between an activity cost and the cost object. The next step is to identify a list of overhead items to study and a list of departments, jobs, and cost codes to apply the overhead to. Many contractors think in terms of assemblies or systems, or work packages as described in previous chapters. Cost objects, or assembles, include field activities such as concrete foundations, structural steel, and rough carpentry. There are many different ways to group or categorize or arrange costs; basically there are not any 'rules,' but some guidelines apply. The activity-based costing process divides cost objects into categories; six activities are too few and 1,000 are too many. Tracking costs on a phone call by phone call basis is too detailed. Pareto's 80-20 rule is helpful here, as it is throughout many aspects of cost accounting and financial management. Costs can be assigned by CSI specification divisions or sections but by assembly is a better application. A contractor should establish an organization and cost hierarchy or levels, similar to Figure 10.1. This would begin with corporate departments, different projects within a department, systems such as site work, substructure, superstructure, finishes, MEP (mechanical, electrical, and plumbing), and finally by direct work assemblies or by foremen or by subcontractors.

The outcome of the workshop, identifying and linking cost drivers to cost objects, is accomplished through a logical series of steps:

1. Identify overhead activities and assign cost of resources, such as indirect labor.
2. Approximate the percentage of time spent on each activity.
3. Multiply time spent by the wage for each person.
4. In the home office, staff personnel must proportion their time according to jobs proposed, jobs bid, construction schedules delivered, site visits, marketing calls, and others. All of these must be measurable and must have distinguishable cause and effects.
5. At the jobsite, management personnel may proportion their time by assigning measurable cost drivers such as requests for information written, meetings attended, meeting notes produced, submittals processed, change order proposals (COPs) processed, and others.
6. Calculate unit rate for each activity: For example, a jobsite project manager spends 20% of his or her time on COPs, at a wage rate of $130,000 is $26,000, at 100 COPS = $260/each (EA), and compare that to the profit received from each COP.
7. Then focus on a) how to redirect overhead and b) reduce overhead or drop activities or focus on other activities, which is the focus of lean construction.

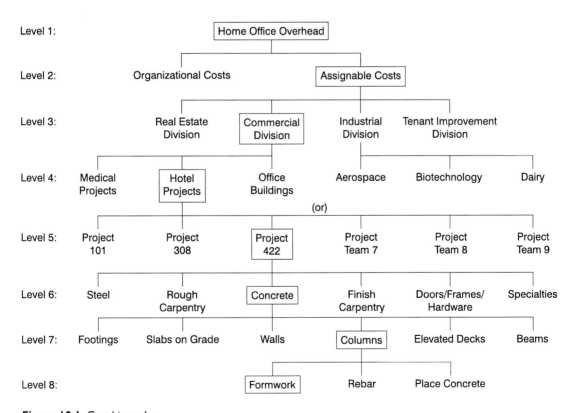

Figure 10.1 Cost hierarchy

Home office activity-based costing applications

The goal of ABC at the home office general conditions level is to first identify the indirect costs and then:

1. Proportion indirect costs according to the divisions which rely on them the most;
2. Attribute home office costs to jobsite;
3. Move construction equipment from the home office ledger:
    a. to jobsite overheads,
    b. to direct work activities, and/or
    c. to subcontractor scopes where possible.

Some of the ways contractors can accommodate activity-based costing are to allocate home office staff personnel proportionally to the projects they are working on and not uniformly charge their time to overhead costs which are distributed evenly across all company operations. This includes not only wages but proportions of office expenditures, office equipment, and automobiles. Some examples of personnel and activities which may be proportioned include:

*   Marketing director or officer,
*   Scheduler,
*   Estimator,
*   Quality control inspector,
*   Safety inspector,
*   Sustainability expert,
*   Cost engineer,
*   General and specialty superintendents including concrete superintendent, structural steel superintendent, earthwork superintendent, and others,
*   Preconstruction department personnel and operations,
*   Senior project managers (see following example),
*   Equipment manager, and others.

Example Two: Each of the major construction company departments shown in Figure 10.1 was headed by a senior project manager (SPM) who provided guidance to project managers and project engineers (PEs), but spent over half of their time marketing their respective market sectors, which included real estate development, biotechnology, medical, aerospace, hotels, and others. They were paid an approximately equal wage of $130,000 per year but the medical market was over 60% of the total company volume, biotech was 15%, and tenant improvement (TI) was hit and miss as shown in Table 10.1. All of their wages were cost-coded to the same generic home office labor cost code. A study of activity-based costing would have first isolated each of the SPM's wages to their departments, and then shifted more of the SPM's marketing time towards the commercial industry, which includes medical and hotels and others, and away from the smaller market sectors such as TI and real estate.

Table 10.1 Overhead spread on divisions and volume

| Column: | 1 | 2 | 3 | 4 | 5 | Method 1 | | | Method 2 | | |
| | | | | | | 6 | 7 | 8 | 9 | 10 | 11 |
Construction Division	Total Volume $ x 1 mil.	Percent of Total V	Fee Earned $ x 1000	Fee Percent of Projects	Fee Percent of Total Fee	HOOH Proportioned by Division	Profit after HOOH	Net Profit Percentage	HOOH Proportioned on Volume	Profit after HOOH	Net Profit Percentage
Real Estate	Nil	0%	$0	0%	0%	$445	−$445	−45.0%	$0	$0	0.0%
Commercial	$75.8	78%	$2,679	3.5%	73%	$445	$2,234	2.9%	$1,392	$1,287	1.7%
Industrial	$18.7	19%	$882	4.7%	24%	$445	$437	2.3%	$344	$538	2.9%
Tenant Improvement	$2.5	3%	$116	4.6%	3%	$445	−$329	−13.2%	$46	$70	2.8%
Totals:	$97.0	100%	$3,677	3.8%	100%	$1,782	$1,895	2.0%	$1,782	$1,895	2.0%
	$ x 1 mil		$ x 1,000			$ x 1,000	$ x 1,000		$ x 1,000	$ x 1,000	

Total Fee as percent of total volume:	3.8%	
Total Home Office Overhead Cost:	$1,781,921	annually
Overhead as percent of total volume:	1.8%	
Net Profit:	$1,895,079	2.0%

Method 1: Overhead spread evenly for each of the four divisions

Method 2: Overhead spread based on each division's volume

A study of activity-based costing is most applicable for construction companies which have a high overhead cost as translated to percentage of revenue and/or high overhead elements such as marketing directors. ABC is also most applicable to companies which have multiple divisions and/or multiple products such that overhead can be directly applied to each division or project as expended and not spread based simply on total revenue percentage. The next two example general contractors practiced ABC and lean construction as a standard course of business.

Example Three: This very large locally owned general construction (GC) firm did a mix of lump sum and negotiated work. Historically they were more cost competitive than other GCs because they employed more crafts direct, such as ironworkers, cement finishers, operating engineers, teamsters, millwrights, and others. They had a substantial warehouse staffed with a full-time warehouse superintendent, two equipment mechanics, and a laborer for cleanup. As the economy changed and they began to employ more subcontractors than direct crafts, there was not much for the warehouse to do so it was eliminated. Without understanding ABC or lean, this contractor employed both of those methodologies successfully.

Example Four: During the 2008 recession, construction slowed to a near halt, especially for private negotiated projects. Many contractors suffered bankruptcy and others incurred significant lay-offs. This construction company survived by eliminating its preconstruction division and vice president of marketing, hired another lump sum estimator, and focused on lump sum competitively bid projects. This was a similar experience to the contractor which moved its office operations all out to one construction project as discussed in Chapter 6, Example Two.

It is important that the process of tracking costs for activity-based costing does not take on a life of its own, costing a company valuable money and resources. It is difficult to track actual costs on an activity-by-activity basis, especially smaller activities or those which constantly change. Capturing true costs is difficult, especially with labor. It is easier, and almost as reliable, to proportion labor cost according to a group of six or so activities. It is easier to proportion home office labor costs than building rent or copy machine rent or furniture. Facility costs can be allocated on department volumes/revenues or on dollars per square foot of floor space they utilize. Labor costs are the most variable and therefore the most applicable to ABC processes. For example, assume the estimator spent 30% of her time on lump sum bid estimates, 60% of her time on negotiated proposals, and 10% maintaining the estimating database, training, and keeping up with new technology. There is some subjectivity in this method but it is still more 'activity-based' than spreading costs simply on revenue. Some indirect costs are easily applied as the next example shows. The same analysis applies for internal non-performing or inexperienced project managers and superintendents who need more support from the home office than do others as shown in one of the examples below.

Example Five: This general contractor offered design-build (DB) services to its client. Many GCs do the same, but without any internal design support. If they land a project, they either contract with a designer or joint-venture. The GC hired an in-house architect to support marketing the DB delivery method. His time would be charged to DB jobs, but when they didn't have any, he wasn't really qualified to do anything else. ABC would therefore likely categorize him as low-value added.

Example Six: Many construction company executives form separate equipment limited liability companies and rent equipment to their own construction companies. This GC assistant controller spent 100% of his time managing these equipment companies and other real estate investments for the company officers, but charged his time to the construction company home office accounting department. ABC would identify him as a no-value contributor to the construction operations and would recommend moving his wages out of company costs. See Chapters 12 and 19 for additional discussions on related but outside activities and investments of construction executives.

Example Seven: This specialty contractor constructs only concrete footings. It has a home office overhead of $300,000 and an average annual revenue of $3 million. They typically place 10,000 cubic yards (CY) of concrete footings in a year at a job cost of $2.7 million, which averages to a very competitive $270/CY. The company only works on one job at a time and employs a field management team of one superintendent and one project manager who are job-costed. Applying ABC to this company results in allocating all home office overhead, both fixed and variable costs, to their job cost because they have only one division and only one product and only one field management team. The $300,000 is added to the $2.7 million to yield a total of $3 million which results in an ABC unit price of $300/CY.

Example Eight: A mid-sized GC performed a mix of bid and negotiated work. This $20 million project for an aerospace client was won on a competitive bid basis, but the client had awarded negotiated work prior and would do so again in the future, so the GC needed to put its best foot forward, even though they had a very tight estimate. The PM assigned to the project had spent ten years as a project engineer and was previously categorized as a career PE. The superintendent also had not worked in this high-stress, high-security environment before. The home office had to send out a senior PM and general superintendent two days a week to keep everyone calmed down and focused and keep the client happy. A revenue-based application of home office overhead would be disproportionally light for this project.

There are many differences in approaches to estimating, cost accounting, and financial management among different construction industries, such as commercial and residential and civil construction as presented throughout this book. There are also no rules among general contractors or specialty contractors for cost control or cost coding or cost allocations. Cost-plus, guaranteed maximum price and lump sum projects are all also accounted for differently. These and other fragmentations in the construction industry make it difficult for researchers to study and apply their findings on new concepts such as activity-based costing.

Jobsite activity-based costing applications

Direct construction activities are work that is self-performed by a contractor and costs are estimated relatively accurately and recorded easily and are not the primary focus of ABC. ABC processes can be applied to many construction materials. For example, a sling of 100 sheets of plywood is purchased by the superintendent with the assistance of the jobsite cost engineer. The plywood, per se, does not have a dedicated cost code, but this sling may be used for concrete forms, site fence, wall sheeting, roof sheeting, floor sheeting, safety signage, a plan table in the office trailer, and other uses. Another example would be five redi-mix concrete trucks with nine cubic yards of concrete in each one, all with the same 3,000 psi (pounds per square inch) design mix. This 45 CY pour day includes footings, sidewalks, and retaining walls. The cost of the concrete should be proportioned, based on CY to each of those separate work assemblies.

Because a general contractor typically subcontracts 80–90% of the work, it is difficult for them to allocate overhead costs to direct work. A pure construction manager who employees 100% subcontractors, which is sometimes a contract requirement, will have a difficult time applying activity-based costing to direct construction cost activities. Poor subcontractor performances cause increased management resources for a GC. This reinforces the need for 'best-value' subcontractors as discussed throughout this book. ABC in this case allows application of home office overhead to projects, but not to direct work activities.

Some indirect jobsite construction costs, or jobsite general conditions, may be applicable to ABC methodology as they are percent of volume based. These costs were described in Chapter 5, 'Jobsite general conditions.' One line item included in the expanded eResource estimate template includes a 15% addition to all equipment rentals for fuel, oil, and maintenance but not proportioned according to any individual piece of equipment or assigned to the work that equipment is assisting with. Another example is temporary power use which can be substantial for projects utilizing a tower crane for hoisting, whereas projects which utilize a crawler crane or boom truck for steel or carpentry will charge the diesel fuel direct to the work assembly and not to the jobsite general conditions budget. Neither of these examples are applicable to ABC. Good examples of jobsite indirect cost allocations to direct work packages include:

- Concrete vibrator rental to cost of foundations or walls or columns,
- Crane and welder for steel erection cost-coded to structural steel,
- Compressor for nail guns to rough carpentry,
- Boom truck to wood framing,
- Personnel hoists to siding,
- Scaffold to exterior envelope,
- Assigning activities and costs to subcontractors, and others.

As discussed throughout this book, construction is unique and fragmented. Each company is different, each project is different, and each jobsite team is different. But within each construction company they should use standard cost codes and cost control systems. Cost coding often stems from CSI and adds a project number and another designator for labor, material, equipment, or subcontractors. A consistent cost coding and accounting system is necessary for an application of activity-based costing.

The future

Activity-based costing success requires buy-in at the top management levels of the construction company, similar to total quality management and partnering. In addition to lean, ABC has an overlap and relationship with other advanced cost control topics such as just-in-time material deliveries and earned value. In order for ABC to become economically feasible it relies on a need for automatic cost recording systems. Leadership within construction companies will recognize that the benefit of applying ABC processes must be greater than the cost of their implementation.

Some new trends in the construction industry have caused both home office and jobsite overhead costs to increase, ultimately with the goal of lowering direct construction costs. Application of these new concepts requires the contractor to invest additional overhead resources in personnel, training, office equipment, office space, and consultants. Some of these trends include:

- Building information modeling,
- Sustainability,
- Lean construction,
- Design-build delivery method,
- Preconstruction services,
- Marketing for negotiated projects versus lump sum bidding,
- Integrated project delivery,
- Construction manager-at-risk, or CM/GC, and others.

All of these concepts have a cost-benefit trade-off, but the costs of implementation are usually charged to overhead. The ultimate goal of ABC is first to identify indirect costs and then allocate them to the cost of work in the hopes of reducing indirect costs. In some regards this may conflict with these new trends. For example, many general contractors estimate that it costs an additional $30,000 to $70,000 for indirect costs per project to achieve LEED certification, especially at the higher gold and platinum levels.

Incorporation of ABC may have a difficult time keeping up with technology changes. Project managers, who used to rely heavily on office support such as estimators and schedulers and clerical staff to type letters, are doing more work autonomously. This reduces home office staff and is a natural method of both ABC and lean. Today most PMs will do their own typing, produce their own estimates, prepare subcontracts and purchase orders, create their own schedules, and process pay requests. PMs who take control of all of these activities are sole-source PMs as was introduced in Chapter 3. Large overhead contractors and those organized around staff project management processes would therefore be more applicable to the benefits of ABC.

ABC research studies have looked at allocation of overhead from in-house contractor fabrication shops to job costs, such as rebar; these are easy to apply. But in most cases contractors purchase fabricated rebar from separate fabrication shops which bid competitively, such that the contractor is already paying market costs and an ABC application is not as effective. Lessons learned from ABC construction material fabrication studies may transfer to costs associated with jobsite material handling and reduction of double-handling inefficiencies; these are all subsets of lean construction.

Increased use of subcontractors reduces risks for general contractors as discussed throughout this book, but causes the company and projects to become more fragmented; this is sometimes a contract requirement. Therefore the future of ABC may be more with subcontractors than with GCs. Subcontractors also fabricate more of their own materials, which is applicable to ABC, such as structural steel, rebar, mechanical ductwork, kitchen cabinets, and others.

Some jobsite general conditions items can be hired out as subcontractors, therefore moving indirect costs from the GCs portion of the estimate to the list of subcontractors. Some GCs items which may be subcontracted include security fence, tower crane, security services, survey, scaffold, and others. This is another natural method of applying ABC.

The benefit of ABC methodology is best suited for contractors which are not already tracking their cost objects sufficiently. Improvements would be most beneficial to contractors in the following order:

1. Contractors which do not separate costs between divisions or projects or indirect from direct costs;
2. Contractors which do not separate labor from material costs;
3. Contractors which combine all types of labor crafts, carpenters and laborers and ironworkers, together including their labor burdens; and
4. Contractors which cost code strictly by CSI and not by assembly.

Summary

The foundations of activity-based costing stem from production industries such as car manufacturing. Many of the lessons learned can be directly applied to the manufacturing of construction materials, but are more difficult at the construction jobsite level. Incorporation of ABC processes in a construction organization starts with a commitment from the company executives. Often an ABC workshop is facilitated for a specific construction company where cost drivers and cost objects are identified.

Home office general conditions are categorized as either fixed costs such as officer salaries, or variable costs such as specialty superintendents. Variable home office GCs are more easily assigned to company divisions, projects, and work activities whereas fixed costs are customarily spread across the entire company based on revenue. In many cases contractors are already practicing a natural form of ABC by moving home office GCs to jobsite GCs and from jobsite GCs to direct work activities. Because contractors typically compete for work where cost is a major component, especially that of jobsite GCs as has been discussed, they are always seeking means of cost savings.

The purpose of ABC is to track and assign as many of the home office and jobsite indirect costs as is possible to the work activities that they directly support. The ABC process begins with assigning costs to divisions within the company, construction projects, and ultimately, if possible, direct construction activities. There are many opportunities and benefits of ABC including identifying inefficiencies and unnecessary costs, specialization of types of projects, and strategic decisions on subcontractor or direct work packages. ABC identifies the more-profitable scopes and allows contractors to assign resources to those scopes and therefore improve profits. There are many ABC impacts to accuracy of future estimates including an improved database through as-built estimates, lowering prices of overpriced items (those which do not require much indirect support), and raising prices of underpriced items (those which require a disproportional amount of indirect support); this should improve pricing competitiveness in the marketplace.

Adoption of lean construction processes makes the identification of cost activities through ABC more applicable. Lean is the focus of the next chapter in this group of cost control and financial management topics.

Review questions

1. List three different terms which mean home office GCs.
2. Does ABC apply to every contractor?
3. Does ABC apply to every construction project?
4. Which types of a) contractors and b) construction projects lend themselves best to ABC processes?
5. If HOOH is reduced, and fee is kept constant, what happens to profit?
6. Of the eight examples presented in this chapter, which a) already have benefited from ABC and/or lean without necessarily identifying it as so, and/or b) could benefit from ABC applications?

Exercises

1. Does the use of ABC require a facilitated workshop? When would this be most applicable and why?
2. Solve for HOOH from the revenue equation and its derivatives earlier, creating one long equation utilizing as many of the variables and cost elements as possible.
3. Review the expanded GCs template on the eResource. What additional items might be moved from jobsite general conditions to direct work activities?
4. This is not necessarily an ABC question, but what are some of the reasons a GC would choose to employ subcontractors versus performing the work with its own direct craftsmen?
5. Table 10.1 has been expanded on the eResource to show which types of work Evergreen Construction Company (ECC) performed with more or less direct labor and which types of work were generally negotiated versus bid. ECC is considering adding an additional home office annual warehouse cost of $220,000 to support their direct work activities and adding a marketing cost of $130,000 to support their negotiated projects. Assume that in the short term, neither of these potential additions effect revenue, but they will increase HOOH and subsequently reduce net profits. Utilize ABC to apply these contemplated additional costs to the divisions which they support most and recalculate the net profits for each.

6. Utilize Figure 10.1 to track a construction system to the eighth level or below, but for a dairy construction project and a different construction system than concrete, such as conveyors or process systems.
7. Other than the examples listed here, provide three examples of a) home office GCs costs, b) jobsite GCs costs, and c) jobsite direct costs.

11

Lean construction techniques

Introduction

Lean construction techniques are a collection of processes intended to eliminate construction waste and still meet or exceed the project owner's expectations. The application of lean principles results in better utilization of resources, especially labor and materials. The strategy for lean supply is to provide materials when needed to reduce variation, eliminate waste, improve workflow, and increase coordination among construction trades. The jobsite project team's responsibility is to create a realistic construction flow that reflects the general contractor's (GC's) and the subcontractors' dependency on material suppliers. Material deliveries are then scheduled so that the materials arrive on site just as they are needed for installation on the project. While lean construction does not replace a construction project's detailed contract schedule, it utilizes short interval planning and control that improves the timely completion of construction tasks. Another aspect of lean construction is the expanded use of off-site construction or prefabrication of building components. Reduced fabrication on the project site enables workers to concentrate on installation of material fabrications rather than creation of construction components, which ultimately saves time, reduces material waste, enhances on-site safety, and improves quality of installation. Acceptance and incorporation of lean for a company or project requires top-down support from the ownership of the project owner, GC, and architect teams, all the way through design engineers and jobsite construction management, to subcontractors and foremen.

Implementation of lean construction is the next logical step from activity-based costing (ABC) and other cost control methods such as work packages and earned value discussed in previous chapters. Similar to ABC, there have been many academic articles and research by Ballard, Howell, Kim, and others on lean construction. In addition to lean techniques, this chapter includes other advanced cost saving and cost reduction methods. This includes target value design, just-in-time (JIT) planning, last planner, pull planning, value engineering (VE), subcontractor and supplier impacts, supply chain material management, and jobsite laydown and material handling. Many of these – such as JIT, last planner, and pull planning – are direct subsets of lean, but this chapter has combined all these advanced cost control topics into one discussion.

Lean construction

There is not one exact definition of lean construction, but rather a body of research leading to a philosophy of planning and implementation of advanced cost control methods. A lot of lean topics are intermixed. Contractors do not need to adopt them all in order to accept a lean philosophy; in fact, some argue that a few of these concepts don't apply strictly to lean. But if 'eliminate waste' is the ultimate goal, and loosely defined, then these all fit under the lean umbrella plus additional advanced cost control concepts; essentially there is no limit. The Lean Construction Institute's website defines lean as "enhances value on projects and uncovers wasted resources such as time, movement, and human potential" (www.leanconstruction.org).

Like activity-based costing and other cost control topics, lean has been adapted to construction from production industries such as the automobile industry, specifically from Toyota. Minimizing waste is not just for construction materials but eliminating any inefficiency such as a surplus of inventory, improving labor productivity, and creating a satisfied client are all premises behind lean. Anything that doesn't add value must be eliminated; lean creates value. Lean is influenced by total quality management where the lack of quality in any facet of design or construction results in rework which disrupts the flow of work. Increasing speed of design or construction in an effort to save money can actually cause waste and increase defects and therefore costs more money. The adage that 'time is money' does not always hold true. Since construction starts with design, the root of many lean savings ideas stems from improving design operations. Although the term *lean construction* is widely understood, it could be expanded to include *lean design* or even *lean built environment.*

Lean topics such as pull planning or just-in-time are not tools, but processes. Lean maximizes value by minimizing waste. Lean is not happenchance; rather it needs a formal process and adoption by all the team members for it to be a success. Acceptance of lean processes also requires a top-down commitment from design and construction company owners as well as project owners or clients. Repeat design- and construction- and project owner-teams are a benefit to the lean process. If they have worked together they know what to expect from one another. Lean is also best suited for complicated projects and ones that use an integrated project delivery (IPD), where all three parties share in the same risks and sign the same contract as shown in Figure 11.1. IPD

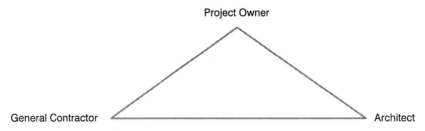

Figure 11.1 IPD organization chart

improves lean success and lean improves IPD success. If IPD is not the project owner's chosen delivery method, then design-build or construction manager-at-risk, also CM/GC, delivery methods would best implement the cost savings topics discussed in this chapter. Lean is also better suited for negotiated projects than bid projects.

The construction industry tends to resist new concepts such as activity-based costing and lean, labeling them production oriented. Construction contractors, and their proven means and methods, are also slow to change as shown in the next example. It is common to hear out on the jobsite that "if it isn't broke, don't fix it." The construction industry is different than other industries as discussed in previous chapters. Some identify that the problem with contractors is their detailed focus on independent activities, estimates, and schedules. It is better to focus on bigger picture goals and interaction of activities. Subcontractors also cause fragmentation in cost analysis and therefore make it more difficult for contractors to adopt lean. Though many construction leaders do have an open mind to improving their processes and have adopted many of the techniques discussed here.

Example One: This large international design-build construction company went through a major transformation during the mid-1980s when it laid-off over half of its engineering work force and replaced very experienced licensed nuclear engineers with younger engineers at a much lower salary. They also increased their hiring of internationally educated engineers, paying them less than the U.S. engineers under the premise that their international degrees were not as acceptable, even though these engineers could operate a calculator and design pipe and wire and concrete just as good as their U.S. counterparts. This new work force struggled with incorporating new energy codes and building codes, especially after the Three Mile Island nuclear incident. Eventually the contractor hired many of the laid-off experienced engineers back, but this time as consultants at twice the wage they had been receiving as full-time employees. Simply reducing or eliminating costs is not always a cost savings move.

The following sections introduce target value design, just-in-time deliveries, last planner, pull planning, and supply chain material management. These concepts and processes are all subsets of lean construction. The term 'lean accounting' is synonymous with lean construction and is the process to document actual cost results from implementing lean construction improvements.

Target value design

Target value design (TVD) is a subset of lean construction. TVD has also been referred to as target value costing and originated in manufacturing industries – again, from the automobile industry. Similar to lean, TVD works well with negotiated and design-build projects. The project owner sets the budget and the design team is tasked to design to that budget. The budget is divided up amongst the different design disciplines like pieces of a pie. Each design package must financially fit within their piece of the pie, so the entire project meets the owner's budget; they design to the target cost. This process either assumes the designers have estimating capabilities in-house or an outside estimating consultant is engaged, or a construction manager is contracted early. Some critical subcontractors may need to be brought on to the team early during design development such as mechanical, electrical, and plumbing (MEP), precast concrete, curtain wall, landscape, elevator, and others. This process then makes it difficult to obtain competitive bid pricing which therefore makes TVD sometimes unfeasible especially on public works bid projects.

The owner's satisfaction is the number one priority for the design and construction team, especially if the owner's focus is on cost. The design team must anticipate cost when making design decisions rather than wait until bids are received. If the budget is exceeded the project must be redesigned, often with input from value engineering. The design must be targeted to 'final' cost as a design parameter, which is often different than bid or contracted cost. The target value design process includes active management of owner contingency logs and value-added logs, which are similar to a subcontract buyout log or VE log.

A target value design project often starts with a workshop where all of the team members buy into the concept. A considerable amount of collaboration is needed between all of the parties, not only the project owner, architect, and general contractor, but also the design engineers and major subcontractors. The estimate is established from the top down, which in construction is the opposite way most estimators work, generally beginning with a more detailed focus, such as quantity take-offs and pricing recaps, and summarizing as they work towards the final price as shown previously in Figure 4.1. In this regard, TVD is similar to pull planning which will be discussed later where the end is determined and the plan is developed in reverse. If one element of the project is over budget, such as structural steel, money must be moved from another system to compensate, such as electrical. Like many lean concepts, TVD connects well with building information modeling. Fourth and fifth dimension modeling can help with early contractor-generated estimates.

Target value design has some pitfalls though. Each of the budget pieces of the pie might initially fit together, but when contractor bids are solicited there is no guarantee that prices will correspond. Project owners often request detailed schedules of values from contractors with their bids and proposals. But contractors have been known to weigh one portion of the estimate heavier than another due to potential change order opportunities or front-loading strategies of early pay request line items. TVD also does not consider changes of economy, such as the price of copper pipe and wire escalation, or supplier and contractor backlog changes which can impact pricing. Change orders also occur on most construction projects and the owner's budget needs to anticipate these. Design-build and CM/GC project deliveries are both understood to have fewer change orders than traditional lump sum bid projects and therefore fit the TVD model better. A satisfied owner is looking for the design to match final cost, not just budget or even bid cost. If the project utilizes a cost-plus pricing method with no guarantee of total price, this places the owner at risk. Although the initial project budget and bid might have fit the target, the final price might not.

Just-in-time deliveries

Just-in-time deliveries of construction materials have been adapted from JIT manufacturing. The goal is that materials which will be needed within 36 hours of installation are on site, but not too early or too late. The contractors perform a balancing act with having just enough material but not too much. Remote projects such as the Olympic Hotel case study will not lend themselves as well to JIT as would a downtown metropolitan skyscraper. Construction sites which are too large may not be managed lean as materials may be spread too far from their place of installation. Too much material on site for too long also exposes it to weather damage and potential theft. Doubling handling material is one of the most significant causes of labor inefficiency as will be discussed later in this chapter. But a problem with holding suppliers off from delivering too soon assumes that the fabricator has room in their storage yard (therefore not lean) and can be stored out of the weather and secured. Other problems can occur when the fabricator sells your project's materials to another general contractor who needs it quicker or is willing to pay a premium and the supplier grabs whichever material is easiest to move or sell.

Last planner

Planning is hard work; it is look-ahead scheduling. The individual or group of individuals responsible for accomplishing the work is labeled 'the last planner.' This includes carpenter and electrician foremen on the construction side, and the structural and electrical engineers on the designers' side. The principle architect is not the last planner. Most architectural firms employ project managers (PMs) just as general contractors, project owners, and subcontractors do. The last planner concept should extend down through the organization chart to the project engineers (PEs), cost engineer, and jobsite cost accountant if possible.

Many general contractors develop top-down schedules and dictate to subcontractors when their work is to be performed. A collaborative schedule created with subcontractor input, not the marketing salesforce but foremen and superintendents, is a much more feasible schedule. The same benefit can be had when a GC's proposed project superintendent assists with developing the estimate. Follows are examples of superintendents which did not utilize the last planner lean construction process.

Example Two: This superintendent was very experienced in concrete tilt-up buildings. He had spent the last ten years working in one office building complex, building one concrete tilt-up shell after another. He did not include his foreman in any planning issues. The foreman did not meet with the architect or owner or even the GC's own project manager. The superintendent would give the foreman hand-drafted sketches he had personally crafted from the drawings, one footing or column at a time, not allowing the foreman to have a set of drawings. Unfortunately, the superintendent suffered a heart attack and could not return to work. The contractor's executives felt that since the foreman had worked for the same superintendent for ten years that he must be ready to step in and run the work. Unfortunately, he failed and was eventually dismissed. If the foreman had been included as the last planner he might have been ready for the increased responsibility.

Example Three: Another general superintendent was building a high-technology facility out of state. He had two trailers, one for him and the GC's management team including a meeting room for the project owner, and another trailer for three assistant superintendents and foremen. These assistants were focused on the carpentry, concrete, and structural steel work. The general superintendent never introduced his assistants to the design or owner teams. He asked that the project manager and project engineer not visit the foremen's trailer as they would disrupt his crew. The general's superintendent pushed the direct work portion of the project, but the project owner's focus was on the MEP subcontractor scopes. Because the PM and PE did not communicate directly with the foremen and assistant superintendents, this portion of the work was not well coordinated and resulted in rework and change orders. Although the concrete was finished under budget, the project owner chose another GC for his next negotiated project. The project owner, GC, and assistant superintendents and foremen could all have benefited from the last planner process.

Pull planning

Pull planning or pull techniques are also adopted for lean construction from manufacturing industries. The people who are the pullers are the ones who use the output and would be considered the 'last planners' as discussed earlier. The process may also begin with a workshop and facilitator. Each stakeholder should be represented, including subcontractors. Just as target value design starts with the total budget and works backwards, pull planning starts with work which needs to be in place in three to four weeks and plans backwards to accomplish that goal. Each contractor and designer is given a set of colored sticky notes and posts their delivery dates and commitments and what needs to happen to accomplish those dates. This would be comparable to each subcontractor developing a three-week look-ahead schedule, which is often done for the weekly foremen's meeting, and then those schedules would be shared with the other companies. But in this case, all of the companies collaboratively develop one schedule together. A photograph of the pull planning schedule from the Olympic Hotel and Resort project is included as Figure 11.2.

Lean construction enthusiasts feel that contractors should not overcommit or overload their schedule. A little under-loading is acceptable as that allows for adjustments during the process if necessary. Conversely many contractors would argue that crews need to be given just slightly more than what they can accomplish in a week, not less. Some contractors feel that employing a superintendent or foreman who is slightly underqualified to perform his or her duties, but is ambitious and wants to work hard and learn, is a better fit than an overqualified individual. This philosophy would support a slight overload of work versus an underload.

The goal of pull planning is not only to achieve collaborative buy-in from all parties, but to work together to remove all constraints from accomplishing the work, so that the only thing left for the last planners is to get the job done. Some of the categories of potential restraints which should be considered include:

- Project owner has funded the project;
- Design is done and has been quality checked and change orders will be limited;
- Permits are in hand and inspectors are scheduled;

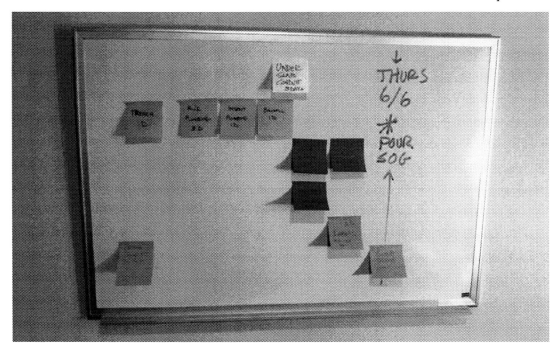

Figure 11.2 Pull planning schedule

- Requests for information and submittals have been processed and answered;
- Material is on site, or will be on site;
- Qualified and sufficient labor is on hand;
- Predecessor activities have all been identified and will be accomplished;
- The jobsite is clean and safe;
- Sufficient time and cost estimates have been allowed for; and
- No rework and no waste are the ultimate goals.

Value engineering

Value engineering to some members of the built environment, especially some designers, implies cheapening of the project's design, but that is not true VE. VE includes analyzing selected building components to seek creative ways of performing the same function as the original components at a lower life-cycle cost without sacrificing reliability, performance, or maintainability. VE studies may be performed by consultants during design development, as a contractor-performed precon-struction service, or by the contractor during construction. The most effective time to conduct such studies is during design development. Target value design, as just discussed, needs to happen before design to be efficient; VE is the opposite and often happens either during the deign process or after design. The traditional lump sum delivery method of design–estimate–VE–redesign results in wasted time and additional design fees and therefore is not considered a lean construction topic by some.

Value engineering studies are typically performed during the preconstruction planning phase. The intent of the process is to select the highest value design components or systems. The essential functions of each component or system are studied to estimate the potential for value improvement. The VE study team needs to understand the rationale used by the designer in developing the design and the assumptions made in establishing design criteria and selecting materials and equipment. The intent of VE is to develop a list of alternative materials or components that might be used. Preliminary cost data is generated, and functional comparisons are made between the alternatives and the design components being studied. The intent is to determine which alternatives will meet the owner's requirements and provide additional value to the completed project. Often life-cycle cost data is analyzed for each VE alternative before presentation to the project owner.

Preparation of value engineering proposals includes analysis of advantages and disadvantages of each alternative and the ones representing the best value are then selected for presentation to the designer and the owner. The VE proposal looks very similar to a post-contract change order proposal, including all detailed costs and markups and substantiation. All cost data prepared by the VE team should be included with the proposal, similar to a change order proposal. Each VE proposal is tracked in a log, similar to other cost control or equipment tracking logs managed by the construction team. The proposals are submitted to the designer and the owner for approval. If approved, VE proposals should be incorporated into the contract documents. VE proposals approved after the construction contract is awarded must be incorporated into the contract by change order. A sample VE log from the Olympic Hotel and Resort case study project is included here as Table 11.1.

Table 11.1 Value engineering log

		Evergreen Construction Company Olympic Hotel and Resort Updated: 2/20/2020					
Item	Description	Date Proposed	Value Proposed	Value Accepted	Value Rejected	Date Resolved	Comments
14	Replace janitor's closet VCT with sheet vinyl	1/15/2020	$3,500	$3,500		1/16/2020	Better maintenance
15	Move window blinds into owner's scope	1/22/2020	−$15,550				Still open, tax considerations
16	Add one more year to landscape warranty	1/22/2020	$5,750		$5,750	1/31/2020	Owner will contract separately
17	Open paint spec to three suppliers	1/31/2020		−$12,322		2/10/2020	
	continued…						

Subcontractors and suppliers

A typical commercial general contractor today subcontracts 80% to 90% of the construction project. Subcontractors, also referred to as specialty contractors, therefore, are important members of the GC's project delivery team and have a significant impact on the GC's success or failure. Since subcontractors have such a great impact on the overall quality, cost, schedule, and safety success for a project, they must be selected carefully and managed efficiently. There must be mutual trust and respect between the GC and the subcontractors, because each can achieve success only by working cooperatively with the other. Consequently, project managers and superintendents find it advantageous to develop and nurture positive, enduring relationships with reliable subcontractors. PMs and superintendents need to treat subcontractors fairly to ensure they remain financially solvent not only to finish this project but to be available to provide competitive bids on future projects.

The use of subcontractors by the general contractor is a risk management process. Subcontractors provide the GC with access to specialized skilled craftsmen and equipment that they may not have in-house. One of the major risks in contracting is accurately forecasting the amount and cost of labor required to build a project. By subcontracting significant segments of work, the GC can transfer much of that cost risk to subcontractors. When the project manager asks a subcontractor for a price to perform that scope of work, the subcontractor bears the risk of properly estimating the labor, material, and equipment costs. Craftspeople experienced in the many specialized trades required for major construction projects are expensive to hire and generally used on a project site only for limited periods of time. It would be cost-prohibitive for a GC to employ all types of skilled trades as a part of their own full-time work force.

Subcontracting is a risk-reduction process, but it does not eliminate all risk. The project manager and superintendent give up some control when working with subcontractors. The scope and terms of the subcontract define the responsibilities of each subcontractor. If some aspect of the work is inadvertently omitted, the general contractor is still responsible for ensuring that contract requirements are achieved. Specialty contractors are required to perform only those tasks that are specifically stated in their subcontract agreements. Consistent quality control may be more difficult with subcontractors as shown in a few of the examples discussed throughout this book. Project owners expect to receive a quality project and hold the GC's superintendent and PM accountable for the quality of all work whether performed by the GC's direct crews or by subcontractors. Subcontractor bankruptcy is another risky aspect of subcontracting, which can be minimized by good prequalification procedures and timely payment for subcontract work. Scheduling subcontractor work often is more difficult than scheduling the GC's crews, because the subcontractor's craftsmen may be pulled off the GC's project and moved to other projects. Many GC superintendents feel safety procedures and practices of subcontractors are not as effective as those used by the GC's own work force. All of these financial and control aspects are even more crucial when subcontractors also subcontract portions of their work to third-tier subcontractors and suppliers as shown in Figure 11.3.

It is imperative that the general contractor selects quality 'best-value' subcontractors if the project manager and superintendent are to produce a quality project, on time, safely, and within budget. PMs must remember that poor subcontractor performance will reflect negatively on their professional reputations and their ability to secure future projects. Once the subcontractors have been selected, contract documents are executed documenting the scopes of work and the terms and conditions of the agreement. Each subcontract and supplier purchase order will receive a unique

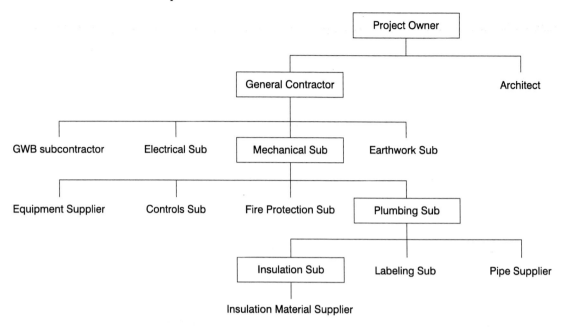

Figure 11.3 Organization chart

cost code and buyout contract values will be input into the cost control system. Subcontractor management is an integral part of project management. While the project superintendent manages the field performance of the subcontractors, the PM and cost engineer manage all subcontract documentation and financial affairs and are responsible for ensuring that subcontractors are treated fairly and paid timely.

This short introduction on procurement and subcontracting has not necessarily been a 'lean' cost control topic. But as introduced in Chapter 10, 'Activity-based costing,' subcontracting has an impact on the general contractor's ability to implement ABC and lean processes. As mentioned, the GC has passed financial risk to its subcontractors, especially if they are contracted lump sum. In one sense using the marketplace to buy out subcontractors and suppliers at the lowest competitive price is a cost reduction tool and is therefore lean but only if additional best-value attributes are considered beyond cost such as quality, schedule, and safety performance. Ignoring the role of subcontractors in any discussion of cost control is omitting 80–90% of the scope of the construction project. This brief discussion of subcontractors and suppliers naturally leads into material management covered in the next two sections.

Supply chain material management

Many lean construction research case studies assume the general contractor is also the material supplier, which is not customary. GCs rely heavily on support from subcontractors and suppliers to successfully build a project as discussed earlier. All construction materials are provided through suppliers. Cost control responsibility within the fabrication facility belongs to the supplier. Good supply chain material management by the GC relies on shifting as much on-site material fabrication and assembly to suppliers, especially local suppliers. Supply chain material management shifts the cost control focus as close to the beginning of a construction project as possible, to design and material fabrication, and not solely on the traditional cost control focus of a contractor's direct jobsite field labor.

Off-site prefabrication

Removing or minimizing construction waste is a fundamental plank of lean construction and is accomplished through prefabrication of construction materials. Off-site prefabrication of materials improves *quality control*, as the materials are built and stored out of the weather, facilitates in-process inspections, improves *safety* in a controlled environment, reduces *cost* as it changes fabrication of many construction materials from project-based to product-based, and improves *schedule* adherence as many activities can be performed in parallel rather than in series. Materials may be specified by the design team as 'purchased' products, rather than field-assembled, but oftentimes a general contractor can weigh the advantages of shop fabricated versus field assembly as part of their means and methods preconstruction planning. There are many prefabrication shops and assembly yards which specialize in assembly and sale of construction materials. In addition to construction of single-family prefabrication homes, commercial construction examples of off-site material prefabrication include:

- Standard examples: Cabinets, structural steel, mechanical duct, concrete rebar, wood trusses, precast concrete beams and concrete planks;
- Occasional examples: Structural insulated panels, apartment wall and floor wood-framed panels, brick panels, bathroom plumbing carriages;
- Potential examples: Whole apartments or hotel units, concrete columns, concrete floors, concrete walls, and others.

Applications of off-site prefabrication extends into activity-based costing, just-in-time planning, building information modeling, and other cutting-edge cost saving construction processes. The following are great examples of efficient use of supply chain material management and off-site prefabrication.

Example Four: The State of Alaska has very cold winters and the typical building season is limited to four, or at the most six, months a year. One general contractor in Juneau, Alaska invested in a large heated warehouse. He fabricates many building components local, in a secure, well equipped, and environmentally controlled facility. This allows him to win many more projects than his competitors as he can limit the amount of time his crew is working in the cold weather. He fabricates almost complete buildings including floor and wall and roof elements and puts them on a barge and ships them out to remote islands for quick cost-efficient assembly. He also attracts the best labor in the area as his crew is offered year-long employment.

Example Five: Timber frame construction is an old system with a new life. It is very similar to post and beam construction. This Pacific Northwest specialty contractor chooses only the best trees and fabricates every beam and column and rafter for pre-assembly in its yard. It pre-drills for oak peg connections and ships the materials ready for assembly to the remote San Juan Islands. The construction system is not the least expensive, but it is local and efficient with excellent quality and has life-cycle benefits.

Example Six: General contractors may purchase door frames from one supplier, door leafs from another supplier, door glass from a third, and door hardware from a fourth. If the door frames and door leafs are a combination of wood and hollow metal, there may be at least two more suppliers. If all of these materials are shipped 'loose' and uncoordinated, there will be substantial assembly labor expended in the field by the GC's carpenters. Many contractors will package all of these elements into one supplier, but that is not always cost efficient. Another method is to have the materials shipped from one vendor to another, including the painter, such that when the doors, frames, and hardware arrive on site they have been 'machined' and are ready for installation without rework, which is a goal of lean.

Local material purchases

It is not only lean to purchase construction materials locally, it is also sustainable and a good means of quality and schedule control. A general contractor which chooses suppliers that fabricate materials locally can easily physically visit the warehouse or yard and witness progress. Third-party inspectors may also visit the shop and inspect and test as is the case with structural steel fabrication. Materials that are purchased and/or fabricated locally, say within 100 miles of the project site, will utilize less fuel in transportation. Purchasing locally results in additional Leadership in Energy and Environmental Design (LEED) points or credits being awarded, if the project owner is pursuing a sustainable certificate or plaque as shown in Figure 11.4. There is also a non-measurable, maybe moral, reason a contractor should do business with as many of its local neighbors as reflected in the next example.

Figure 11.4 LEED plaque

LEED® and its related logo is a trademark owned by the U.S. Green Building Council® and is used with permission (usgbc.org/LEED)

Example Seven: This small general contractor built quality custom and speculative homes in his rural county. He had a great reputation and never had to compete for work. Even though he was a residential GC, he was union and only employed union carpenters and laborers and cement finishers and subcontractors. He believed strongly in 'trading local' which to him meant you bought your clothes from the only clothing store in town and didn't travel to the 'big city.' The people he traded with would in turn employ him to build their homes and commercial building projects. He purchased his gas for his personal cars and business equipment from the local gas station, even though it was a little more expensive than another station 15 miles away; he had built the business owner's home as well. The gas station owner later hired a contractor from out of town to build an investment apartment project. Consequently, the local residential GC did not purchase any gas from this station again.

Jobsite laydown and material handling

It is imperative that the jobsite be organized efficiently in order for the construction project to be successful. The jobsite layout affects the cost of material handling, labor, and the use of major equipment by the general contractor and the subcontractors. A well-organized and clean site has a positive effect on the productivity of the entire project work force and their safety. The jobsite layout plan should identify locations for temporary facilities, material movement, material storage, and material handling equipment. Ignoring this may have an adverse effect on productivity, safety, and cleanup as reflected in the next example. The proper choice of tower crane types and locations is essential for a productive plan, especially on downtown hi-rise projects. The Olympic Hotel and Resort project fortunately had ten acres of site with plenty of laydown area and access for mobile cranes. The site layout plan should be developed by the GC's superintendent and it should consider not only the needs of the GC's direct work, but also the requirements of all of the subcontractors working on site. The site plan should indicate:

- Existing site conditions, such as streets, adjacent buildings, and overhead and underground utilities;
- Planned locations for all temporary facilities such as fences and gates, trailers, temporary utilities, sanitary facilities, erosion control, and drainage; and
- Material handling and storage areas, equipment storage and access, craftsmen and visitor parking, and hoisting methods such as a tower crane.

Example Eight: Every construction site element needs to be considered when developing the site logistics plan. Every construction project is visited by a vendor truck, also known as a food truck or taco truck and some other fun nicknames. Some jobsites have multiple trucks showing up before work, again at the 10:00 am coffee break, and at noon for lunch. The vendor trucks need to have the superintendent's approval with respect to location and timing and cleanup. This GC superintendent did not approve of vendor trucks; he felt they resulted in wasted time for his craftsmen and generated garbage. His crew and all of the subcontractors would break for coffee and lunch and walk two blocks to a corner convenience store. This distance extended their breaks away from work. On the way back to the site occasionally some garbage would end up on the ground and the superintendent would send a laborer out for four hours a week at the end of shift on Friday to clean up. Keeping the truck off the site to save labor actually cost more labor.

The superintendent should consider all site constraints, equipment constraints, jobsite productivity, material handling, and safety issues when preparing the jobsite layout plan. Some additional factors that should be considered in developing the jobsite plan include traffic flow on and off the site and location of site offices convenient for visitor control. Some of the goals of the site logistics plan are to: eliminate double handling of material which is a lean productivity factor, protect materials from weather and theft, and keep materials close to work but not too close such that the craftsmen are stumbling over them. These are all just-in-time lean considerations as discussed earlier. In addition to a productivity tool, the jobsite layout plan is a great proposal, interview, and marketing tool. It shows the project owner that the contractor has thought through the project – a personal touch. It may make a difference on a close award decision with a negotiated project. The jobsite layout plan for the Olympic Hotel and Resort is shown on the eResource.

Summary

Lean construction loosely defined includes designing, fabricating, and constructing projects as cost effectively as possible, eliminating waste, all the while maintaining the same if not improved levels of quality, schedule, and safety control. Lean construction has several sub-categories including target value design, just-in-time material deliveries, last planner participation, and pull planning scheduling processes.

Target value design starts with the project owner setting a strict budget that both the design team and the construction team must adhere to. Construction projects, such as skyscrapers, have been practicing just-in-time material deliveries before JIT was included as a study of lean construction. Efficient JIT management requires delivery of materials when needed, not too early so that double handling is minimized and not too late such that it affects the schedule of installation.

Ultimately the people involved in designing and installing materials, the last planners, should be the ones involved in estimating, scheduling, and cost control. Essentially this is the structural engineer who sizes the spot footing and the carpenter foreman responsible for excavation, forming, and placing the footing concrete. Pull planning is a collaborative scheduling approach which includes those last planners preparing short interval schedules by scheduling backwards, often with multi-colored sticky-notes.

Value engineering seeks the same goal as target value design but happens generally after some design has been completed. Contractors prepare cost-saving or life-cycle-saving proposals for project owners and designers to consider and accept, which then require re-design and incorporation into the general contractor's contract and various subcontract agreements. Subcontractors comprise 80–90% of the work force on any typical construction project. True lean construction savings should focus on subcontractors and suppliers.

Supply chain material management re-focuses cost control from on-site labor to the source of material design and fabrication. This is a more holistic approach to lean construction and cost control, starting at the beginning and chasing down material sources and cost of fabrication. Shifting site labor to off-site warehouses and supplier fabrication yards saves field labor and potentially improves quality, safety, and schedule control. Some examples of material which benefit from off-site prefabrication include structural components such as wood and steel trusses, concrete beams and columns, and finish materials such as cabinetry and door assemblies. Off-site prefabrication also reinforces just-in-time planning. It is not only good for business and community economies to purchase materials from local material suppliers and employ local subcontractors, but it is good for the environment due to reduced fossil fuel use in material deliveries. Purchasing local also allows the GC and project owner and design team to visit fabrication shops and perform early quality control reviews. Sourcing locally also has safety enhancement considerations and employs JIT material planning. Local material purchasing employs many lean construction techniques.

A preconstruction jobsite laydown plan should be developed by the project superintendent. This plan, which includes many of the lean construction planning processes discussed in this chapter, is a physical drawing of the project site. The superintendent will include many cost, schedule, quality, and safety enhancement considerations such as material laydown, storm water control, traffic routes, and others. All of these are advanced cost control methods and should be considered part of the lean construction philosophy. There are many more as well, and others yet to be discovered and implemented by future construction cost engineers.

Review questions

1. What is the difference between lean construction and ABC?
2. What is the difference between TVD and VE?
3. List three advantages of buying local materials.
4. What is the difference between a subcontractor and a supplier?
5. What is an advantage of having too much material on site versus not enough material?
6. If room is not available to store material on site, where can it be stored?
7. What happens if the doors, door frames, and door hardware package is bid higher than what was budgeted on a TVD project?
8. List three advantages of off-site prefabrication versus on-site casting for a system such as concrete beams.
9. Other than a hospital, list three project types which may be suited for lean construction processes.

Exercises

1. Other than the examples listed earlier, what types of materials or material systems might be prefabricated off site?

2. Prepare an argument why it would be better to a) only schedule your foreman for 90% of his or her capacity, and/or conversely b) schedule your foreman for 110% of his or her capacity.
3. Are there any of the eight examples included in this chapter that you disagree are lean construction related? It is okay to disagree, just prepare your argument or make a suggestion for how it could be related.
4. Do you know of any other lean cost control techniques?
5. Prepare three VE ideas from your case study (hotel or other) and include them in a VE log.

12

Equipment use and depreciation

Introduction

When choosing topics to be covered in this book we talked with contractors and accounting firms and researched other construction accounting books. One topic that is very important to the concept of cost accounting has always been depreciation. Depreciation is defined as the *wear and tear on fixed assets* such as real estate, construction equipment, and office furniture. Depreciation is an important means of reducing taxes for investors and corporations, which is especially true for construction companies which own equipment. But the process of accounting for depreciation is not necessarily a project manager's responsibility; tracking and allocating depreciation is more a chief financial officer (CFO) and home office accounting function. Similar to other markups on the bottom of the estimate summary such as fees and insurance and labor burden, accounting for depreciation on the equipment ledger has usually been accomplished by the home office accounting department, not the jobsite project management team. This is often done internally through use of accounting journal entries. If construction companies, not equipment companies, own their own equipment, the accounting department may create an invoice and send out to the jobsite to have the project manager (PM) cost code and approve the rental costs. Depreciation reflects negatively on the financial statements as an expense, but is a paper expense only, not a reduction in cost.

All along, the approach to this book has been to connect home office accounting with other financial responsibilities of the jobsite team including the project manager and the jobsite cost accountant. This chapter therefore expands on the discussion of construction equipment depreciation to include additional related jobsite project management concepts, as well as defining types of construction equipment and their ownership structures; internal or external. If the equipment is owned outside, the project management team will need to arrange and manage equipment purchase order contracts with equipment companies to secure rental. If owned internally, ideally all construction equipment will be job costed and not sitting in the contractor's storage yard. Equipment requiring maintenance needs to quickly be repaired to support jobsite productivity and the cost of those repairs needs to be accounted for in the proper location. Estimating and cost coding of construction equipment rental and repairs was introduced in previous chapters.

The rules for accounting of depreciation are determined by the Internal Revenue Service (IRS) as it relates to corporate income taxes and more specifically allowance for tax deductions. Taxes and depreciation are complicated topics and ones that the contractor will rely on expert accounting help for, possibly with the services of an outside certified public accountant (CPA).

Equipment types

How many different types of construction equipment are there? There is likely close to an infinite list of different types of equipment and it would be difficult to cover each one. Tools are differentiated from equipment for this example as tools would be purchased for a project and are expendable, but equipment might be rented and depreciated and should last longer than for just one project. There are several good construction equipment textbooks available, such as J. Schaufelberger's *Construction Equipment Management* (1999) which describes equipment operation. Following are just a few basic lists of equipment which the construction team might encounter:

Heavy-civil contractors:

- Graders,
- Dozers or bulldozers,
- Dump trucks,
- Track hoes or excavators,
- Compactors,
- Scrapers, and others.

Marine contractors:

- Barges and Tug boats,
- Marine cranes,
- Pile drivers,
- Under-water equipment, including welders, and others.

Cranes:

- Tower cranes:
  - Horizontal jib tower crane, also known as 'hammer-head crane,'
  - Luffing tower crane,
  - Mobile tower crane, also known as self-erecting or 'fast cranes';

- Mobile cranes, including trawler or crawler cranes;
- Light cranes, including truck cranes also known as 'cherry-pickers' or boom trucks;
- Miscellaneous cranes not typically used for construction:
  ○ Gantry cranes,
  ○ Ship yard cranes,
  ○ Bridge cranes;
- Personnel and material hoists and temporary elevators are not cranes per se, but their estimating and operation considerations have many similarities to tower cranes.

Commercial and residential construction equipment:

- Forklifts,
- Pickup trucks,
- Flatbed trucks,
- Backhoes,
- Scissor lifts,
- Scaffolding is not necessarily considered equipment but the accounting for rental and erection and maintenance is similar to other methods of equipment.

Light equipment:

- Surveying equipment, including total-stations and lasers,
- Compressors,
- Welding machines,
- Construction electrical generators,
- Concrete vibrators and troweling machines and screeds, and others.

Office furniture and equipment:

- Computers and printers and scanners and plotters,
- Copy machines,
- Desks and chairs and file cabinets,
- Construction office trailers and tool vans and dry shacks are not necessarily considered equipment but are handled similar to many of these other items with respect to accounting.

Equipment ownership

All of these types of equipment may be owned by the construction company and most discussions on construction equipment accounting and depreciation are focused on internally owned equipment. But actually, very few construction companies own any equipment. The exception might be large heavy-civil contractors and marine contractors. They rely on a much larger percentage of specialized self-operated equipment for their construction operations than do commercial and residential contractors which rely on direct labor and the use of subcontractors. Commercial and residential construction is 80% or more subcontract costs whereas civil and marine work is 50% or greater equipment costs. Most commercial and residential contractors will either set up separate equipment companies or will rent from outside sources. Equipment may also be provided and operated by subcontractors as part of their scope and obligations. There are advantages and disadvantages of all of these arrangements.

Internally owned

Contractors that own and operate their own construction equipment endeavor to keep that equipment busy on jobsites whenever possible. Upon purchase, a piece of equipment will simultaneously show up as an asset on the contractor's balance sheet in the amount of its book value, and as a liability in the amount of the loan balance owed to the bank. As the loan is gradually paid off, the liability will be reduced and the amount it has been depreciated will correspondingly reduce its value on the asset side of the balance sheet. These types of contractors are heavily leveraged and purchase equipment through the use of long-term loans which require significant monthly payments. Construction company loans will be discussed as a subsection of the last chapter of this book. If the equipment is not in use and in the contractor's storage yard, loan payments must still be made to the bank. The equipment will continue to age and depreciate and corresponding deductions in value will be made to the corporate balance sheet. When an internally owned piece of equipment is sold, it will also have an effect on the contractor's balance sheet and income statement.

Maintenance and repair costs for internally owned equipment will be charged to jobs if the equipment is in use or charged to home office overhead if the equipment is idle. The accounting for this type of equipment is often performed by the corporate CFO through the use of internal journal entries. Contractors which own equipment must keep it busy or bear significant financial risk. Conversely contractors may choose to own equipment if their equipment requirements are unique and they do not want to be subject to availability and rental charges from outside third-party supply firms. Contractors which own equipment have the flexibility to adjust their own internal rental charges which may allow them to be the low bidder on projects which are heavily equipment dependent, such as civil and marine work.

Contractor equipment companies

To clients and even many on the contractor's team, internally owned versus separate contractor equipment companies may look identical, but they are actually significantly different with respect to accounting and risk management. Many construction companies choose not to own their own equipment but rather set up separate equipment companies or separate divisions, which own the equipment. These separate companies are formed as limited liability corporations (LLCs) and will have a different name from the construction company. The equity ownership of these companies is usually a select group of individuals within the construction company, often the corporate officers. For example, Evergreen Construction Company (ECC) has five executive officers which comprise their board of directors (BOD). Those five officers include the chief executive and operational officers (CEO and COO), CFO, Vice President of Marketing, and Human Resources Officer. These five individuals establish a separate LLC which will own and rent equipment to their own construction company, ECC. They named their equipment company Main Street Equipment, LLC. They are therefore renting equipment to themselves. They charge rental to the construction company very similar to rental charges from an outside equipment company. If there is ever a problem, such as a market downturn or a safety accident which can be attributed to a piece of equipment or operation, then the construction company is financially protected as they had rented the equipment from Main Street. The corporate officers are also personally protected as they formed a separate LLC. The LLC can report a loss on their income statement or declare bankruptcy and shield the equity owners personally, and also the construction company, from a lawsuit or economic harm.

Experienced project owners which utilize open-book contracts, such as the AIA A102, do not see the financial separation between the construction company and the construction company's equipment division that the contractors will represent. They often regard all equipment supplied by the contractor as internally owned and strive to protect themselves from excessive equipment rental charges or maintenance costs. The best method for them to do this is to contractually modify or insert language into the A102 contract Articles 7.5.2 and 8.1 such as:

- Contractor's monthly, weekly, or daily rental rates will not be in excess of the average rates charged by the three largest local equipment suppliers;
- Contractor's cumulative rental on any owned piece of equipment will not exceed 85% of the purchase price (book value would be better but that would be difficult for the client to determine);
- Equipment repairs and maintenance of contractor-owned equipment will not be job-costed.

Outside ownership

The easiest method to manage equipment rental from a project manager's perspective is from outside third-party suppliers which own and rent construction equipment as a main source of business. This equipment is expected to be delivered to the jobsite in perfect condition and is expected to remain that way during the course of construction. Equipment breakdowns have a significant effect on jobsite labor productivity. Contractors can obtain bids from outside sources and negotiate rental rates and conditions. Conversely, the PM cannot really negotiate with his or her own construction company BOD on the rental of one of their own forklifts. The contractor will enter into a purchase order agreement for the rental or lease of equipment which is not furnished with an operator. An example rental purchase order for a forklift is included here as Figure 12.1. If equipment is rented with an operator, a subcontract agreement should be developed by the jobsite team rather than a purchase order. Any company that provides labor on the jobsite must also provide general liability insurance and the subcontract agreement allows for this.

Evergreen Construction Company
1449 Columbia Avenue
Seattle, WA 98202
206.557.4222

Purchase Order

Vender:	Cedar Rentals				PO No.:	2209
	992 Eastlake Ave.				Date:	03/01/2019
	Seattle, WA 98101				Job Name:	Olympic Hotel

| Attention: | Janet Jones | | | | Job Address: | Rt. 4, Box 787 |
| Phone No.: | 206.233.9922 | | | | | Rainforest, WA 98101 |

Item	Quantity	Units	Description	Unit Price	Amount
1	1	EA	50' Extended reach forklift rental	$3950/mo	$3950/mo
2			Minimum 6 month rental period		
3			Supplier services once per month		
4			Supplier responsible for maintenance		
				Total:	$3950/mo

| | Delivery: | 1 week |
| | Shipment: | Your Truck |

Terms and Conditions:

1 Supplier accepts PO issued only by ECC authorized personnel
2 Supplier shall invoice each item separately each month on the 20th of the month
3 Invoices received without PO attached will not be processed

Requested by:	Approved by:	Purchased by:
Randy Smith	*Chris Anderson*	*Jennifer Thompson*
Superintendent	Project Manager	Cost Engineer

Figure 12.1 Purchase order

A very important aspect to equipment rental is the management of economical rental durations. This again is easier to control with outside rental companies than with internally owned equipment. Equipment rental often is accompanied by mobilization or delivery charges and demobilization or pickup charges. These charges can be substantial and repeated rotation of the same piece of equipment on and off the jobsite can add up. The jobsite accountant or cost engineer should work closely with the superintendent to ensure efficient equipment use. There are also more economical durations to rent equipment, for example it is more economical and potentially less expensive to rent a welding machine for:

- One whole day rather than six hours,
- For one whole week rather than four days,
- For one whole month rather than three weeks, and
- For one whole year rather than ten months.

Table 12.1 Rental equipment rates

	Cedar Rentals 992 Eastlake Ave. Seattle, WA 98101		
	Rental Equipment Rates		
Description	Daily ($)	Weekly ($)	Monthly ($)
Air Compressor, 185CFM DSL	118	450	1,075
Crane, Truck Mounted, 14-16T	495	1,750	5,150
Vibration Plate, Medium	70	280	650
Dump Truck, 5YD	325	1,300	3,250
Water Truck, 2,500 Gal	325	1,300	3,900
Forklift, 40'–42', 6,000#, 4WD	285	1,090	2,600
Forklift, 50', 8,000#, 4WD	350	1,300	3,950
Forklift, 55', 10,000#, 4WD	425	1,615	4,620
Generator, 45KVA (32KW)	175	600	1,800
Loader, Skid, Medium	185	705	1,750
Excavator, Mini, 7,500#	235	905	2,580
Backhoe, 4WD, Extendahoe	265	1,010	2,900
Boomlift, 40'–42', ST 4WD	350	965	2,300
Boomlift, 60', ST 4WD	425	1,215	3,465
Boomlift, 80' w/5' Jib, 4×4	865	3,295	7,850
Scissor Lift, 19' Vertical	105	300	760
Scissor Lift, 26' Vertical	110	420	1,000
Scissor Lift, 32' Narrow	140	525	1,250
Trash Pump, 3"	65	190	545
Shopvac, Wet/Dry	33	90	230
Welder, 225Amp, Gas	50	150	465

A very good local Pacific Northwest equipment supplier's partial rental rate sheet is included here as Table 12.1. We appreciate their contributions to this body of work. It is easy to see the economics of managing rental periods from this table as discussed earlier.

Subcontractor ownership

The most cost-effective way for a general contractor (GC) to have equipment on the project is not to rent it at all, either internally or externally, and require all of its subcontractors to provide their own equipment. If a GC operates more as a construction manager (CM) and provides little to no direct labor, he or she may also be able to avoid providing any equipment, or at least minimal equipment. The CM will still provide a trailer for the on-site staff, a pickup truck for their superintendent, office furniture and office equipment, potentially a tower crane and personnel and material hoist, and likely a forklift. But the more equipment is provided by subcontractors, the less the GC or CM needs to worry about idle time, maintenance and repairs, or mobilization costs for equipment.

Equipment operation

Although robots have garnered an increased role in construction today, they are not yet operating heavy construction equipment. How the equipment is operated, by whom, and how it is contracted are important aspects of construction cost accounting and risk management. There are a variety of methods for how operation of equipment is handled. This section briefly introduces self-operation, subcontractor-operated, and supplier provided, subcontractor-operated and provided, and owner-operator equipment options.

Self-operated

If contractors own their equipment, either as part of the construction company or through a separate but internally owned equipment company, they will likely operate the equipment with their own forces. This provides the contractor with the comfort that they know the equipment operator and his or her capabilities and that those individuals are loyal to the construction company. The contractor will pay the operator as they would any other construction craft employee. A time sheet will be filled out by the foreman and then cost-coded to either the equipment code they are operating or the work they are performing as discussed in the 'Equipment allocation to job cost' section later. Labor burden will also be figured for the employee and charged to the job.

Equipment operation is performed by operating engineers (OEs), which are an organized craft and labor union. It is an interesting side note that surveyors are also members of the OE's union. Merit shop, or open shop contractors do not need to employ OEs and may choose to operate equipment with general laborers or construction foremen. Regardless, in most cases, equipment operators must be certified or licensed to operate the equipment they are using. Everyone may be able to drive the jobsite pickup truck, but not everyone may operate the forklift or the tower crane. This author's son is an assistant commercial construction superintendent who is a certified forklift operator and helps with this on occasion, and also spends a couple of hours each day in his pickup truck tracking down materials.

Rented and subcontractor-operated

If equipment is rented from an outside supplier, the contractor may operate it with their own operating engineer as discussed earlier, or they may employ a separate contractor to operate the equipment. This may be the case with large equipment such as a tower crane, where the contractor does not have an experienced or available OE in house, nor wants to take the chance on employing an operator off the street. There are equipment companies which will rent equipment to general contractors with operators (then requiring a subcontract agreement), or rent the equipment bare without an operator (then requiring a purchase order), or furnish the operator only (also requiring a subcontract agreement). If the GC subcontracts out operation-only of equipment, they will pay a loaded wage rate which will include the OE's direct wages, labor burden, and profit and overhead and other markups. The subcontracting firm providing the OE will take care of distribution of wages and contributions to labor taxes such as social security. Some equipment may also require two operators, such as a large crawler crane. One of the operators is in the cab of the crane and the other is attending to maintenance. This second OE, nicknamed the 'oiler,' may also drive the crane when relocating from jobsite to jobsite. Tower cranes will require a separate 'rigger' who may be employed on a subcontract basis as well or employed internally. The rigger is often a member of the ironworkers' union as they are customarily well-versed in crane operation and hoisting and associated safety issues.

Subcontractor provided

The general contractor will not need to concern itself with choice of equipment operators when subcontractors provide their own equipment. An example of this is an earthwork subcontractor that owns and operates a variety of construction equipment as noted earlier with heavy-civil such as excavators and dozers. It is the subcontractor's responsibility to choose qualified employees and handle their wage compensation and labor burden contributions. The subcontractor is also responsible for all equipment licensing and maintenance expenses. The GC will still make sure that the subcontractor is utilizing only qualified equipment operators. Copies of operator licenses and certifications will be made available to the GC either as part of the subcontractor's project-specific safety plan or as a separate submittal.

Owner-operators

Many single pieces of equipment are owned and operated by one individual. This is very common with many forms of earthwork equipment including backhoes, track hoes, and dump trucks. Many long-haul semi-trucks are driven by owner-operators. The general contractor will then pay the equipment owner an hourly (or weekly or monthly) rate which combines the rental of the equipment along with the wages of the owner, his or her necessary labor burden, and markups such as insurance and fee and overhead and licenses. This loaded rental rate will also include costs for necessary maintenance and repairs. For the GC who employs an owner-operator it is similar to one-stop-shopping.

Equipment allocation to job cost

Previous general conditions chapters reinforced a subset of the revenue equation where fee equals overhead and profit. Jobsite overhead is part of job cost. The overhead included with fee is home office overhead (HOOH). Assuming fee is fixed, any reduction in HOOH will improve profits and any increase in HOOH reduces profit. Construction corporate executives will reinforce to the project teams their need to produce an acceptable fee, if not improve upon it. The allocation of equipment costs, including rental and operation and maintenance, plays a very important role in managing HOOH and improving profits. Essentially the construction company wants all equipment to be job-costed and not sitting idle in the storage yard or under repair while away from a construction project. Some of the *strategies* or methods to keep equipment costs from eroding profits are described in the following. Many of these are applications of activity-based costing as discussed prior.

- Move equipment rental from the HOOH to project overhead.
- Move equipment rental from jobsite overhead to job cost; this is moving it from an indirect expense to a direct expense. An example incudes a welding machine and mobile crane which should be charged to structural steel erection.
- Require subcontractors to provide their own equipment, such as forklifts for material unloading and cranes for hoisting to the roof.
- Subcontract more work and self-perform less work, thereby decreasing the need for rented and operated equipment.
- Charge all owned-equipment maintenance to the job and not the home office, especially on negotiated projects where equipment maintenance may be a job cost.
- Include maintenance responsibility clauses in all rented equipment purchase orders or subcontract agreements.

   The jobsite accountant or cost engineer and superintendent will need to work together closely to track equipment which is on the project. There are several project management tools they will utilize to accomplish this. Equipment internally owned by the construction company will be tracked on its *equipment ledger* by the CFO, which includes allocations of rent and maintenance and depreciation. The equipment ledger was introduced in Chapter 7 with other financial statements. On the jobsite the construction team will utilize a simple Excel spreadsheet as an *equipment log* which lists each piece of equipment, its source, cost code, arrival date, departure date, and potentially any comments associated with operation and maintenance.

   Another paper tool used by all general contractors, and most subcontractors, is the *superintendent's diary*. This is also known as a daily report or log or journal and should be authored by the project superintendent. It records a variety of contemporaneous information about the happenings of the project for each day worked. The diary is filled out by the superintendent at the end of each work day. Some of the items it includes are:

- Weather,
- Work accomplished and other work that was hindered and why,
- Material and equipment deliveries and equipment pickups,
- Visitors and inspections,
- Subcontractors working that day, manpower on site, hours worked, and a variety of other important project management reporting issues.

   The diary plays an important role in tracking rental equipment as well. The cost engineer records on the equipment log what equipment is on site and what it is working on, but because of the authenticity of the superintendent's diary it is also reviewed when there is a dispute with a supplier or subcontractor regarding manpower, equipment deliveries, equipment problems, and equipment demobilization.

Equipment rented externally is accounted for very similar to construction materials. Monthly invoices are received and cost-coded by the jobsite accountant and turned in to the main office for payment. Internally owned equipment may be completely accounted for by the main office through the use of journal entries, or equipment time sheets are sent to the jobsite for cost coding by the project financial team. Equipment rental, similar to construction material, is not subjected to retention by the general contractor of the supplier.

Maintenance expense allocations

The fact is construction equipment requires ongoing and routine maintenance and on occasion, breaks down. These costs can be very substantial and are not typically anticipated or allowed for in a contractor's estimate. Looking at the detailed general conditions estimate for the hotel case study project on the eResource, there was a line item for equipment maintenance but only a minor cost allowance had been included. Even more expensive than maintenance are the cost ramifications to construction productivity when the equipment is shut down. Imagine working on a 40-story skyscraper and the tower crane or the material and personnel hoist are shut down for a week! The best method a contractor can use to manage these costs is to only rent equipment which is in very good condition, require the provider to perform routine maintenance and pay for all repair costs, and include a clause in the purchase order requiring the supplier to have a backup machine ready in case there is a complete breakdown and replacement is necessary.

Equipment which is owned by the contractor, whether internally or through a separate but internally owned equipment company, will endeavor to have its maintenance costs paid for at the jobsite and not cost-coded to the home office. The best method to accomplish this is to repair the equipment while it is still on the jobsite. While under repair, the jobsite team may continue to pay rent for its own company's equipment, even though it is not in use, again to keep costs off of the home office books. Even when the equipment has served its use on the project, it may stay on the jobsite equipment log for an extra week and routine maintenance is performed before it is shipped to a new project or the company's storage yard. Savvy open-book clients may not allow contractors to job-cost equipment maintenance, which increases the contractor's risk of self-owned equipment. An additional risk of self-owned equipment is ongoing safety code or regulation changes dictated by the State which require contractors to upgrade or modify their equipment. If the equipment is externally rented these risks are then all the supplier's responsibility.

Depreciation

Depreciation reflects the wear and tear or normal aging that fixed assets such as buildings and construction equipment realize. Assets lose value immediately after purchase; they all have a useful or productive life and don't last forever. Although the exact duration of the useful life of any asset is variable, the IRS has established a limit of depreciable years for each type of asset which is discussed later. The pickup truck the superintendent drives is not worth the same in the second year of operation as when he or she first drove it off the lot; it has 'depreciated' in value. If only rent is received on assets, and depreciation is ignored, at the end of the useful life of the equipment there would not have been a fund readied to purchase its replacement. One purpose of depreciation is to create a reserve fund for equipment replacement.

Only fixed assets can be depreciated. This includes buildings, construction equipment, furniture, and office equipment. Bare land and people cannot be depreciated. Most construction companies are not in the business of buying and improving or building and operating buildings. This is the purpose of real estate developers as will be discussed later in this book. Real estate is a very risky business and if the construction company does own buildings, they will likely organize separate LLCs for each piece of property, similar to equipment companies. This discussion will focus primarily on depreciation related to contractor-owned construction equipment. When equipment is purchased it represents an asset on the balance sheet as well as a liability either in increased owner's equity or a long-term bank loan. The asset and liability balance out.

Most of this discussion on depreciation is focused on company-owned equipment and buildings. If contractors rent equipment from outside sources, or if subcontractors provide equipment, then they will resolve their depreciation issues similar to maintenance and licenses discussed earlier. The management of company-owned equipment depreciation processes are CFO related and not typically jobsite accounting related. Although companies which own a lot of equipment will attempt to have at least their depreciation and loan payments covered by jobsite rent or revenue, so it is worthwhile for the project manager and jobsite cost accountant to understand some of the basics behind these processes. There are three primary reasons to track or account for depreciation:

- Equipment revenue and cost and depreciation affects financial statements;
- Equipment costs are allocated to project costs in lieu of home office overhead; and
- Tax reporting and payments.

The total and monthly portion of the original purchase value that the contractor plans to depreciate is determined by subtracting the expected salvage value from the purchase value and spreading over its useful life, as is reflected by the following equations:

> Purchase value – Expected salvage value = Total depreciation available
> Total depreciation / Useful life = Depreciation per year

Depreciation for tax purposes

Depreciation represents one of the largest tax deductions available not only to construction contractors which own buildings and equipment, but to any corporation or investor. Depreciation represents a loss in asset value and therefore a loss in income, on paper, which is used to offset other financial gains. Depreciation is a paper loss of the asset, there is not an actual outflow of cash such as is the case with wages paid to construction craftsmen or the purchase of concrete. Corporate accountants and CFOs who prepare taxes, or their outside CPAs who prepare taxes

Table 12.2 Depreciation recovery periods

Asset Example	Useful Life (years)
Office computers and copy machines	5
Most construction equipment, including forklifts and pickup trucks	5
Other office equipment and furniture	7
Barges and Tug boats used in marine construction	10
Qualified leasehold improvements	15
Farm buildings	20
Water treatment plants	25
Residential real estate	27.5
Non-residential real estate	39
Railroads and tunnels	50

on behalf of contractors, will follow strict depreciation guidelines set out by the IRS. The IRS does not make tax laws, Congress creates the laws. The IRS manages the tax collection process passed by Congress on behalf of the U.S. Government. Two IRS publications related to construction asset depreciation include Form 4562, and its instructions, and Publication 946 (www.irs. gov). The useful lives for assets as determined by the IRS are included with these instructions. Table 12.2 includes IRS-determined useful lives for some of those items which are most common for construction companies.

The IRS recognizes three systems available to contractors to depreciate their equipment for tax purposes. The three methods for tax depreciation include straight line and two methods allowed by law under the modified accelerated cost recovery system (MACRS). These two are based on the concept that an asset loses more of its value during the early years of its life and less towards the end. These two include the 200% double declining balance approach for those assets with a useful life of five years or less and the 150% declining balance approach for equipment with a useful life of greater than five years.

The formulas and variables utilized for depreciation are briefly introduced here and discussed again later. The depreciation rate (R) and depreciation value (D) formulas here reflect the straight-line depreciation method, where equal amounts are deducted each year over the use life of the asset:

- P: Initial purchase price of an asset
- S: The expected future salvage value
- N: The number of years allowed for recovery as previously listed in Table 12.2
- R: The depreciation rate per year in percentage; $R = 1/N$
- D: The depreciation value for a given year (n); $D_n = (P-S)/N$
- B: Book value which is the current value of an asset calculated by subtracting cumulative depreciation from original purchase price; $B = P - \sum D$

The depreciation amounts for each year for a piece of equipment with a useful life of five years under the *straight-line depreciation method* would be accounted for as follows:

Year one: 20%; Year two: 20%: Year three: 20%; Year four: 20%; and Year five 20%

The *double declining balance (200%)* and the *150% declining balance methods* allow for accelerated depreciation in an asset's early years. Most construction companies utilize the 200% method for maximum early-year tax deductions. The straight-line depreciation rate equation of R = 1/N is replaced with R = 2.0/N and R = 1.5/N respectively. These two accelerated methods also replace (P–S) in the straight-line depreciation value equation with the previous year's book value, such that Dn = (Bn–1) x R. Five years of depreciation for a typical piece of construction equipment would therefore be subject to the following depreciation rates utilizing either the double declining balance or 150% methods:

200%: Year one: 40%; Year two: 24%; Year three: 14.4%;
Year four: 8.6%; and Year five: 5.2%*

150%: Year one: 30%, Year two: 21%, Year three: 14.7%,
Year four: 10.3%, and Year 5: 7.2%*

Note (*): Each of these methods allows for more depreciation than allowed if an adjustment is not made during the last year. The total which can be depreciated cannot exceed the book value (B), which is the original purchase price (P) less the expected salvage value (S). Therefore in the above declining balance example equations, the calculated depreciable amount reflects 5.2% and 7.2% of the book value, but the actual tax deduction will likely be less and a manual calculation will be necessary. Assets with depreciable useful lives other than five years will have different depreciation rate percentages.

Two additional depreciation methods, known as Section 170 Deduction and Bonus Depreciation, allow for even greater early year tax deductions than the double declining balance and 150% methods, but we will leave that discussion to the CPA and CFO board rooms for the present.

All three of the depreciation examples assume a purchase date of January 1, which is unlikely. Very few assets will be purchased on January 1 of a given year. Full-year depreciation for the first and subsequently the last year of the equipment's life is therefore less than the middle years. The IRS also allows for half-year, mid-quarter, and mid-month depreciation conventions. If a piece of equipment with a five-year life was purchased on January 1, and it utilized the straight-line approach, it would have five equal portions of depreciation. If it were purchased on July 1, it would receive 50% of a full year's depreciation in the first year, equal full-year portions in years two through five, and a second 50% deduction in the sixth year; this being the mid-year approach to depreciation. A mid-quarter purchase divides the calendar into eighths and allows for a 1/8th portion in the first year for the quarter it was purchased, assuming it was purchased in February, 25% equal portions in the remaining first year quarters (totaling 7/8 for the first year), full year proportions for years two through five, and in the sixth year the final 1/8th deduction. Numerous other combinations are available for purchase during other months and quarters. A similar, more-complicated approach is available if a contractor chooses the mid-month convention where the year is divided into 24 segments. Table 12.3 reflects straight-line depreciation examples of each

Table 12.3 Straight-line depreciation options

			Forklift Example, Five-Year Useful Life						
			Purchase Value – Resale Value = Depreciable Value						
			$125,000 – $25,000 = $100,000 = $20,000 per year						
Date	Depreciation	First Year							
Purchased	Method	Portion	2019	2020	2021	2022	2023	2024	Total
1/1/2019	Full Year	100%	$20,000	$20,000	$20,000	$20,000	$20,000		$100,000
7/1/2019	Half Year	50%	$10,000	$20,000	$20,000	$20,000	$20,000	$10,000	$100,000
2/15/2019	Mid Quarter	7/8	$17,500	$20,000	$20,000	$20,000	$20,000	$2,500	$100,000
4/15/2019	Mid Month	17/24	$14,167	$20,000	$20,000	$20,000	$20,000	$5,833	$100,000

of these methods for the purchase of a $125,000 forklift with a useful life of five years and an expected resale value of $25,000.

The sale of company-owned equipment also affects the balance sheet and the income statement. The book value of the equipment is theoretically what it is worth at any point in time. The book value stems from the original purchase value less accumulated depreciation and is reflected in the equation following. The book value is therefore the expected salvage value at a given point of time. If the equipment sells for more than the current book value, the company realizes a profit and must pay income tax on the sale. That profit will be taxed as a long-term capital gain and will be subject to a flat 15% or 20% tax (depending on the individual or firm's marginal tax bracket, see Chapter 18) which is usually less than what the company pays on construction profits. If the equipment is sold for less than the book value, the company realizes a loss and can either use that loss on their current income taxes to offset other gains or roll the loss forward to a year when there are other gains available to offset.

Purchase value – Accumulated depreciation = Book value
Sale value – Book value = Profit or Loss

Depreciation for internal accounting purposes

For equipment which is owned by a contractor and fully job-costed, it is not a negative cash flow draw. But equipment which is stored and not in use does reflect negatively on the contractor's income statement. Internally owned equipment is also not a source of profit for the contractor, as the amount of rent paid by the job, likely from a journal-entry transaction, is approximately equal to the amount the equipment has depreciated or lost value and loan coverage. Internally owned and job-costed equipment is subject to the following financial statement actions:

- Balance sheet: The book value of the asset is reduced each month by the amount it is depreciated. The loan on the asset, which is a long-term liability, is correspondingly reduced and funded by the rent received.

- Income statement: Equipment rent and maintenance expenditures are paid as job-cost and should equal invoices charged to and revenue received from the client. Company-owned equipment records depreciation as an expense on the income statement.
- Equipment ledger: Value is decreased or depreciated, and a reserve fund is increased.
- Job cost history report: An ongoing and complete to-date history of rent paid by the job and maintenance costs if applicable.

If equipment is idle and not job-costed there is a totally different set of financial statement results, including a loss in potential revenue for the company. The goal of company-owned equipment is to be 100% job-costed. The accounting for equipment is different for commercial contractors than for heavy-civil contractors which tend to employ fewer subcontractors and own and operate more of their own equipment. If equipment is not company-owned, then expenses are noted on the income statement and handling of invoices is similar to material purchases and reflected on the job cost history report and the accounts payable report.

There are three methods available to contractors to internally account for depreciation including straight-line and two accelerated approaches, the sum-of-years or digits, and the declining balance approach. The straight-line and declining balance approaches are similar to those utilized for tax purposes. The *sum-of-years depreciation method* recognizes accelerated depreciation in the early years of an asset. The straight-line depreciation formula is revised for this method as $D_n = (P-S) \times$ fraction. The fraction varies each year depending on the length of the useful life of the asset and the amount of years remaining to be depreciated. The fraction, shown as a percentage here, corresponds approximately to the following rates for a typical piece of construction equipment with a useful life of five years:

Year one: 33.3%; Year two: 26.7%; Year three: 20%;
Year four: 13.3%; and Year five: 6.7%

Summary

There are many elements of construction equipment which warrant careful consideration by the contractor's ownership and the project management team, including the jobsite cost accountant. There are a variety of different types of equipment and assets which must be managed and accounted for, including depreciation for financial planning and tax purposes. Contractors can own equipment, they can set up separate equipment companies as LLCs, or they can rent equipment from outside sources. Contractors do not want construction equipment sitting idle in the company's storage yard; they endeavor to have all equipment active on construction jobsites. One way to manage costs and shift risks is to require subcontractors to provide their own equipment. All construction equipment must be operated by individuals who are trained and certified. Maintenance and repair of equipment can have substantial cost ramifications as well as jobsite productivity impacts. The project manager must look to his or her contract with the client to establish requirements and allowances for equipment rental and repairs, especially with respect to contractor-owned equipment.

Depreciation is defined as a loss in value of an asset over time. All fixed assets owned by a construction company can be depreciated, including buildings, equipment, and office furniture. The IRS has established three methods and durations a contractor can utilize on their tax returns for depreciation of each asset. The straight-line approach assumes an equal depreciation amount each year over the life of the asset. Two accelerated depreciation methods account for equipment which loses more value early in its life, thereby allowing the contractor to generate larger early tax deductions.

Review questions

1. How can a client keep the contractor from charging excessive rent or maintenance costs for a company-owned piece of equipment?
2. If a civil contractor which owned 30 pieces of major earthmoving equipment had to store all of that equipment during an economic downturn, how would that equipment be cost-coded?
3. If there are so many risks with owning equipment, why would a BOD of a commercial construction company set up a separate LLC and get into that business?
4. How do third-party equipment rental companies account for their equipment repair and maintenance?
5. Looking back to our Table 12.3 forklift depreciation example, assuming a 2019 purchase date, what is the book value of the equipment at the end of the third full year of ownership for each of the options presented?
6. What is the difference between book value and salvage value?
7. If a PM for a GC is provided with a company car, where is the rental of that car charged: a) on an open-book negotiated project, and/or b) on a closed-book lump sum bid project?
8. How would your answers change to Question Seven if the PM was working a) out of the home office, and/or b) assigned to the jobsite?
9. What are two uses this book has for the abbreviation 'OE'?

Exercises

1. Other than the uses mentioned earlier for the superintendent's diary, what else might be included on that important project management tool?
2. Provide an example of at least one additional piece of equipment which could be added to each category of the equipment types listed in this chapter.
3. Provide two examples of why the jobsite team should rent equipment from an arm of the construction company and two why they should rent from outside sources.
4. If you were a PM, would you rather obtain your construction equipment from an outside third-party equipment supplier, rent from a separate construction equipment division of your employer, or use internally owned equipment? Explain your answer.
5. How might you as a jobsite accountant legally get around the limitations the client places on your owned equipment charges in an open-book project? Is your answer ethical?
6. Prepare a spreadsheet for your CEO assuming you are a jobsite accountant or a PM comparing all of the different options for equipment ownership and operation and maintenance responsibilities. How are they contracted? Who pays for what?

7. Draw an organizational chart for a construction firm which rents a tower crane from a supplier, hires a moving firm to deliver it to the jobsite and return it to the supplier when complete, hires another firm to erect it and later dismantle it (including the use of an 'assist crane' and operators and ironworker crew), hires a crane operator from the original supplier (but not direct – rather through the supplier), and hires a firm to visit the jobsite once a month and inspect the crane and perform routine maintenance. Include your superintendent, foreman, and rigger on the organization chart.

8. Of all of the entities mentioned in Exercise Seven, which should be given a purchase order, which should be subcontracted, and which should be paid as direct employees?

9. On January 1 of 2017 your company purchased the following equipment. How much will you depreciate for the fiscal year ending in 2020 when you have your CPA file your tax return on April 15, 2021 assuming a straight-line depreciation method? How much will you have depreciated total as of this point in time? What are the remaining book values for each piece of equipment?

- Dump truck, $100,000, expected resale value of $20,000
- Backhoe, $280,000, resale $40,000
- Five computers, $1,000 each, no resale value
- Flatbed trailer, $75,000, resale $10,000
- Five pickup trucks, $30,000 each, $2,500 resale value each
- Two construction jobsite trailers, $20,000 each, $5,000 resale value each

13

Cash flow

Introduction

There are a variety of tools that are used in construction. Architects use computers today, not drafting boards, and produce drawings with software such as computer-aided design (CAD) and building information modeling. Surveyors used to use a level or transit but today they most likely use a laser or total-station. Carpenters use hammers and plumbers use pipe wrenches in the field. Project managers (PMs) and project engineers (PEs) use a variety of construction management document tools as discussed throughout this book, most of them prepared and transmitted on the computer as well. Accountants use various financial statements as tools including balance sheets, income statements, and cost ledgers. They no longer rely on large tablets of grid paper but produce most of their accounting documents also on the computer.

Even though many construction management tools have changed, there is one tool that project owners have used with general contractors (GCs) and GCs likewise with subcontractors for hundreds of years, and will continue to do so into the future, and that is *cash*. Project owners also have the next potential contract as a tool to use with GCs, and GCs do the same with subcontractors, but without cash, contractors will realize financial difficulties and potentially suffer bankruptcy. Cash for this discussion is not necessarily actual dollar bills and coins but a positive flow of money through the bank. In fact, very few contractors will deal in hard currency and those that do may be trying to avoid various taxes. Even contractors who report profits, have a good reputation for quality work performed safely, and bring their projects in on time may still have financial difficulties if they are not being paid on time and do not have a positive cash flow. The lack of a good positive flow of cash has an even greater effect on subcontractors and suppliers, especially those which are subcontractors to subcontractors and are far removed from the client and the bank. The lack of cash flow is likely the single most common reason for contractor failures.

The focus throughout this book has been on the accounting and financial efforts of the jobsite team. But cash flow is an important focus of the contractor's chief financial officer (CFO) and chief executive officer (CEO), along with other stakeholders including boards of directors and equity partners. Each jobsite is an individual revenue base; that is one of the differences between construction and other industries as has been discussed prior. The home office relies on its construction projects to bring in cash in the form of monthly payments from their clients, and add together all of the project revenues and expenditures to analyze corporate cash flow positions. Many CFOs will use a positive cash flow generated by the construction teams to produce other income in the form of short-term investments such as stocks and bonds, equipment company operations, and real estate investments. The home office cash flow position is beyond the responsibility of an individual project manager and jobsite accountant, but each of their jobsite cash flow efforts contribute to the corporate bottom line.

This chapter will discuss in detail the process of preparing cash flow curves, which requires first the creation of a cost loaded schedule. The concept of cash flow has several elements; most of them included in the broad concept of cash outflow and cash inflow. All of the different jobsite expenditures including labor, material, equipment, subcontractors, and indirect costs have a negative impact on cash flow and require tracking. The only significant positive flow of cash for general contractors is revenue received from the client. Contractors always want to operate in the black so that their inflow of cash is greater than their outflow. There are various methods a contractor can use to improve its cash position and those are discussed here as well. Some of them are ethical and some of them are not.

Cash flow curve process

A cash flow curve is a projection of the total value of the work to be completed each month during the construction of the project. It is created by cost loading the schedule and plotting the total monthly costs. Often, it is one of the first things the owner will ask of the project manager and may be required by the construction contract. One reason this is required is to provide information to the bank for anticipated monthly payments. Some PMs resist on the basis that the curve will be wrong and that they may be penalized for it. The most important thing a PM does is to get paid from the owner for the work that has been completed on the project. This will be discussed in more detail in Chapter 14, 'Payment requests'. If a cash flow curve is a requirement to facilitate payment, it should be developed.

Cost loaded schedule

The cash flow curve is easy to prepare and begins with the development of a cost loaded schedule. Development of the cost loaded schedule by the estimator or cost engineer starts with a summary schedule and a summary estimate. Detailed versions of these may be helpful to prepare the cost loaded schedule but schedules and estimates that are overly detailed with hundreds or thousands of line items might be cumbersome. Schedules and summary estimates with less than 25 activities are probably too few and 40–50 activities would be ideal. Anything with over 100 would still be usable, but potentially unnecessary. The list and description of the activities on both the summary schedule and the summary estimate should more or less be similar. The best method to explain development of a cost loaded schedule is with a set of step-by-step instructions as follows:

1. Start with an Excel spreadsheet. List direct work activities down the left side of the sheet. Add the cost of those activities in the next column. Add a subtotal row below the direct work activities and add the costs vertically down. Verify that this subtotal cost matches the subtotal cost from the summary estimate. Some of this may be cut and pasted from the summary schedule or pay request schedule of values (SOV).

2. Across the top of the Excel spreadsheet list the months from the construction schedule or weeks for a short duration project.

3. Take the estimated costs for each direct work activity and spread them according to when the activities will be complete. Following are four examples of how direct costs might be spread:

    a. If the foundations are worth $90,000 and will be spent in May, then put $90,000 in May adjacent to the foundation line.

    b. Structural steel installation, estimated at $270,000, occurs in months six and seven and can be split evenly at $135,000 for each month.

    c. Assume the exterior skin is worth $720,000 and will start in the middle of December and work continues through January and February and overlaps into March. It is suggested that the costs be pro-rated as:

$$16.67\% - 33.3\% - 33.3\% - 16.67\%$$

    This approximately equates to $120,000 for December, $240,000 in January and December each, and the last $120,000 in March. Exact estimates and dates are not necessary as this is not an exact science. It may be easier and just as accurate at the end of the day to round all these percentages and dollars to the next whole digit. Do not use cents in any of these calculations.

    d. Ten percent of the $1.2 million plumbing subcontract is attributed to month two for under-slab rough-in, 50% occurs over a three-month span when wall and ceiling rough-in are scheduled, and the balance during the last couple of months of the project for trim and testing. This proportions out as:

$$\$120,000 - \$200,000 - \$200,000 - \$200,000 - \$240,000 - \$240,000$$

    If there is a slight adjustment to be made for any of the work line items, do so in the last month.

4. Total the spread of direct work items down for each month.

5. Add a row at the bottom of the sheet below the direct work subtotal row for jobsite general conditions. The general conditions may be spread or proportioned in one of three different fashions:

    a. Evenly spread the general conditions the same amount for each month.

    b. Proportionately spread the general conditions such that if the general conditions amount to 6.8% of the total direct work estimate, attribute 6.8% of the subtotal direct work tally for general conditions across the page.

       Options 'a' and 'b' are both easy to compute and easy for a client to understand and accept, but neither will be completely accurate.

    c. Calculate approximately how much of the jobsite general conditions will be spent for each month. Most projects realize more general conditions at the front end and the back end

Table 13.1 Cost loaded schedule

Evergreen Construction Company

Olympic Hotel and Resort, Project #422

COST LOADED SCHEDULE

3/31/2019

$ X 1000:

Item	Description	Cost	Qtr 2, 2019	Qtr 3, 2019	Qtr 4, 2019	Qtr 1, 2020	Qtr 2, 2020	Qtr 3, 2020	Totals
		$4,826,612	3000	1827					$4,826,612
3	Concrete	$150,000				150			$150,000
4	Masonry	$411,103		411					$411,103
5	Structural and Misc. Metals	$2,223,633		1000	1224				$2,223,633
6.1	Rough Carpentry	$895,560					896		$895,560
6.2	Finish Carpentry	$182,257				82	100		$182,257
7.1	Insulation	$573,638			200	374			$573,638
7.2	Roof and Accessories	$484,333			100	200	184		$484,333
7.3	Waterproofing	$154,586				100	55		$154,586
7.4	Sheetmetal and Flashing	$390,720			150	150	91		$390,720
8.1	Doors and Frames	$720,920			200	521			$720,920
8.2	Windows and Storefront	$173,300			75	75	23		$173,300
8.3	Door Hardware	$1,283,000				500	783		$1,283,000
9.1	Drywall	$390,723				150	150	91	$390,723
9.2	Painting	$663,343					400	263	$663,343
9.3	Floor Covering	$157,085					57	100	$157,085
10	Specialties	$355,237				100	155	100	$355,237
11	Equipment and Furnishings	$435,000				100	200	35	$435,000
14	Elevator	$442,700	100			100	200		$442,700
21	Fire Protection				100	143	200		$442,700

	Description							Total
22	Plumbing			400	400	300	111	$1,211,379
23	HVAC and Controls			100	300	300	21	$720,629
26	Electrical Systems	200	200	200	300	300	87	$1,286,702
27	Low-Voltage Electrical					200	133	$332,500
31	Excavation and Backfill	186	100					$285,500
32.1	Paving					206		$205,550
32.2	Walks and Misc. Site work					241		$240,990
32.3	Landscaping					75		$75,000
33	Site Utilities	300			200	121		$621,271
	Subtotal Direct Costs:	$3,786	$3,538	$2,749	$3,845	$5,037	$940	$19,893,271
	Labor Burden (Spread Evenly)	179	179	179	179	179	179	$1,074,287
	Jobsite General Conditions (Even)	277	277	277	277	277	277	$1,662,783
	Subtotal Direct and GCs:	$4,242	$3,994	$3,205	$4,301	$5,493	$1,396	$22,630,341
	Fee and % Markups (Proportionate)	356	335	269	361	461	117	$1,899,674
	QTR Total, Excluding Tax:	$4,598	$4,329	$3,474	$4,662	$5,954	$1,514	$24,530,015
	Cumulative Totals:	$4,598	$8,927	$12,401	$17,062	$23,016	$24,530,015	$24,530,015

of the project, due to activities such as mobilization, buyout, and close-out, and a more even spread during the middle of the project. This is subjective and difficult to forecast accurately as well.

6. There cannot be too many sets of subtotals to keep the costs straight. Add a column on the far right-hand side of the schedule and add each of the direct work activities across the sheet. This set of totals should equal the original estimates from the far left-hand side of the sheet which were brought forward from the summary estimate. If there is a mistake, correct this now. If slight adjustments are necessary, make them either in the first or the last month that an activity occurs.

7. Add another subtotal row below the spread of general conditions and add the total direct costs to general conditions.

8. Add another row (or more) for markups. All the markups can be grouped together such as labor burden, fee, insurance, contingency, taxes, and others. Calculate the percentage all these markups are of the subtotal for direct and indirect costs. Use this percentage to pro-rate the markups across the sheet to the right. Labor burden may be split out separately on projects which have a large amount of direct labor which often occurs early in the project. Most clients will accept a pro-rata share of markups be invoiced the same way this schedule is being prepared.

9. Add a total row below the markups and add the subtotal direct and indirect costs to the markups. This total in the column on the far left-hand side of the sheet and that on the far right-hand side of the sheet should equal the contract total. If they aren't exactly, go back and correct the error.

The cost data that is now summed at the bottom of the schedule for each month should reflect the anticipated monthly project expenses. Most of the scheduling software programs can prepare an 'exact' schedule of values with input of the detailed estimate, but again the line items must be exactly coordinated. The computer will not do the logical thinking associated with spreading the estimated costs, but rather just perform the math. The likelihood of the general contractor being billed by each material supplier and subcontractor according to any anticipated schedule is somewhat remote. The contractor would not normally provide the client or the bank all the detail on this cost loaded schedule; rather they would list just the monthly total figures from the bottom row. These monthly totals may be adjusted for retention and sales tax as will be discussed in the next chapter on pay requests. Table 13.1 shows Evergreen Construction Company's cost loaded schedule for the Olympic Hotel and Resort case study. Due to space limitations this is a condensed quarterly version of the 16-month analysis which is included with the eResource.

Cash flow curve

Now a cash flow curve can be simply plotted. Again, this can be as simple as a keystroke or two with scheduling software. The cash flow curve is displayed in either bell shape or 'S' shape. The bell-shaped curve represents a plot of the estimated value of work to be completed each month. The S-shaped curve represents a plot of the cumulative value of work completed each month. The best solution is to plot both curves on the same sheet but provide different vertical scales; otherwise the monthly curve is quite flat and does not accurately communicate the change in forecasted cash needs. Some project managers will adjust the monthly figures to reflect a standard or 'perfect' bell-shaped curve. Within reason, this is acceptable for presentation purposes but not a requirement. Many cash flow curves actually depict more of a double-hump camel than a bell. This is caused when there are significant project costs early on in the project, such as pre-payment for long-lead equipment and site and structural work, and significant costs late in the project, such as expensive finishes and mechanical and electrical trim. The example case study is a remote hotel project and significant expenses were realized early in the project for concrete in the building and garage and structural framing and a peak late in the project during interior finishes. The actual cash flow can later be tracked against this schedule. Figure 13.1 shows the contractor's work-in-place cash flow curve derived from the Table 13.1 cost loaded schedule.

One interesting twist to the cash flow analysis is that there are several different means of plotting and measurement:

- Committed costs: The purchase orders (POs) have been issued and the subcontracts have been awarded. Therefore, the general contractor and project owner have committed to spend the money, but it has not yet been paid. This curve is shifted far left of the one drawn in Figure 13.1.
- On-site materials: The reinforcement steel was delivered to the site, but the contractor has not received an invoice for it, and therefore, not made payment.
- Costs in place: The light fixtures that were delivered last month are installed this month. Costs for materials are not usually accounted for until the materials are installed on the project, especially if installed by a subcontractor.
- Costs billed to the GC: This reflects invoices received from suppliers and subcontractors. It usually lags and is plotted to the right of costs scheduled to be in place.
- Monthly pay request to the owner: This follows receipt of invoices from suppliers and subcontractors. The payment request can be up to 30 days after some of the labor was paid and materials were received.

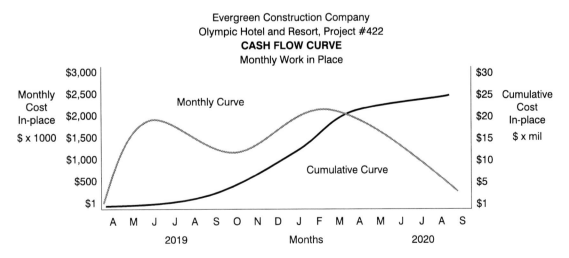

Figure 13.1 Cash flow curve

- Payment received: This reflects when payment was received by the GC from the owner and the owner from the bank. It will lag the formal monthly payment request by 10 to 30 days.
- Payment distributed: This curve reflects payments distributed to subcontractors and suppliers. It will generally lag payment received from the owner by an additional 10 to 30 days. This curve is shifted far right of the one drawn in Figure 13.1.

So, which measure of cash flow should the curve represent? In most instances, the general contractor's project manager will develop the curve based upon his or her schedule of construction that reflects the anticipated costs in place. This was the method used to develop the curve shown in Figure 13.1. The formal invoice will be one to four weeks behind this curve, and the receipt of cash and subsequent disbursement to subcontractors up to a total of two months behind the time when the work was accomplished. In this way, the actual cash flow will fall behind the original projection. The curve, once developed and submitted, should be monitored monthly as another check that the initial project cost control and financial management plan is relatively followed. The preparation of a whole project labor curve presented in Chapter 8 followed a similar process to the development of the cash flow curve.

Jobsite expenditures

As discussed prior, tracking job costs is one of the major and most time-consuming aspects of cost control. Much of the efforts in tracking jobsite expenditures during construction will be the responsibility of the project engineer, cost engineer, or jobsite cost accountant. In order for the fourth step in the cost control process – modification of processes, systems, and potentially people – to be effective, costs needed to be accurately tracked and reported prior. It is suggested the reader refer back to the cost control cycle in Figure 8.1. The processing of buying out subcontractors and suppliers and drafting and executing purchase orders and subcontract agreements was also introduced in previous chapters. Costs are 'committed' when contracts are let for materials and subcontractors and/or when materials are received, and labor is performed on the jobsite. These same committed costs are recorded when invoices and time sheets have been received and/or approved by the jobsite team. The costs are actually expended when checks are cut or electronic deposits are made. Different accounting systems and different contractors may account for 'costs incurred' under any of these scenarios. In this section, each of the activities and steps associated with tracking direct labor, materials, equipment, subcontract costs, and jobsite general conditions are stepped through.

Direct labor costs

1. Labor activities occur during week one.
2. On Friday of week one, the foreman responsible for each craft (carpenters versus laborers versus ironworkers) will fill out a time sheet for his or her crew. Sometimes this is done on a daily basis, especially for larger projects. On that time sheet they report hours worked against a description of work activities and the cost codes assigned to those activities. The jobsite cost accountant may assist the foreman with assigning cost codes. See Table 13.2 for a time sheet example.

Table 13.2 Daily time sheet

	Evergreen Construction Company					
	Daily Timesheet					
Project Number:	422			Project:	Olympic	
Superintendent:	Randy Smith					
Foreman:	Joe Wallace			Date:	12/15/2019	
	Construction Activities:	Grout Frames	Install Frames	Hang Doors	Hardware	
Crew/Craft:	Cost Codes:	8.01	8.11	8.15	8.51	Total Hours
Joe Wallace/Fore	Carp			2	6	8
Ron Roberts	Carp		8			8
Bruce Grun	Carp		4	4		8
Mark Ramstad	Carp			6	2	8
Kelli Gordon	Carp	2		8		10
Rob Murnen	Laborer	8	1	1		10
	Total Hours:	10	13	21	8	52

Overtime Explanation: *Kelly and Rob came in early to get a jump start on grouting the frames before the crew arrived*

Approvals:

Randy Smith   *Jennifer Thompson*

    Superintendent   Cost Engineer

3. The home office accounting department will enter the hours worked, against the cost codes assigned, into the weekly labor report which will be incorporated into the job cost history report. This is primarily a data processing activity.
4. Thursday evening or early Friday morning of week two the home office accounting department will cut checks for the direct crafts for week one. This may be performed by a payroll clerk. Those checks are distributed to the field Friday morning.
5. The general contractor's superintendent, assistant superintendent, or foremen will distribute last week's checks to the craftsmen at the end of day on Friday of week two. Today it is also common for electronic deposits to be made to construction craftsmen's bank accounts.

Direct material costs

1. Materials are received on site throughout month one.
2. Material deliveries will be accompanied by a 'delivery ticket' from the supplier. The foreman or project engineer who received and inspected the material before unloading will sign the ticket. The supplier will keep a copy and the jobsite will retain a copy.
3. The jobsite may enter the delivery ticket into a cost log for material at the site and send the receipt ticket to the main office awaiting the supplier's invoice. The cost code assigned to those materials from the original purchase order should be noted on the delivery ticket.

4. Suppliers send invoices to the contractor's home office. The supplier should note their PO number and their total PO value on the invoice. A copy of the material delivery ticket should be attached to their invoice. If short-form POs were used and the supplier has a copy of the PO, they should attach that to their invoice as well to assist processing.

5. The home office accounting department will match up all of the invoice paperwork from the supplier with their copies of delivery tickets and POs, if they have them, note the material invoice value in an accounts payable log, and send the invoice package to the jobsite for approval.

6. The jobsite cost accountant will receive the invoice and confirm receipt of the construction material, the amount of the invoice, the cost code, and other pertinent information. He or she will then initial approval and forward to the project manager and superintendent for their initials as well. A jobsite inspection may be necessary to validate the quantity of the material on site. On some open-book cost-plus projects the client's on-site representative may initial approval as well.

7. There is no need to deduct retention from suppliers or direct labor or equipment or indirect costs – only subcontractors receive a retention deduction.

8. The approved invoice will be returned to the home office awaiting payment and entered into the job cost history report.

9. If these materials were ordered via a short-form PO, the invoice may be processed within a week or ten days after jobsite approval.

10. If these materials were ordered via a long-form PO, and their costs more significant, payment may be held until month two, ten days after receipt of payment from the client.

Equipment expenditures

1. The processing of equipment rental invoices will be very similar to that of materials ordered with short-form purchase orders.

2. Equipment rentals will be checked against the jobsite equipment ledger and potentially the GC's superintendent's diary if there are disputes regarding time used on the project. Reference the previous chapter for additional discussion on equipment use and depreciation.

Subcontractor invoices

1. Subcontractors perform work throughout the month and incur labor and material and equipment costs similar to a general contractor.

2. Subcontractors are required to submit their pay requests to the GC on the 20th of month one which will forecast the amount of work anticipated to be completed through the end of that month.

3. Subcontractor pay requests sent to the GC's home office will be logged in as accounts payable and forwarded to the jobsite for approval.

4. The jobsite cost accountant will verify that the total amount billed does not exceed the amount approved in each contract and assign the appropriate cost code for each subcontractor. Similar to material invoices, the cost engineer will initial approval on the subcontractor pay requests.

5. The project manager will then review subcontractor pay requests, verify accuracy, and assemble them into the total project pay request to the client for month one. Five or ten percent retention will be withheld from each subcontractor; the same rate the client will be holding from the GC. See the next chapter on pay requests for additional discussion on this process.

6. The PM and superintendent may need to walk the project and verify percentage completion for each subcontractor request. The superintendent should initial invoice approval as well. Many contractors will include additional levels of invoice approvals from home office officers as a means of accounting checks and balances.

7. The PM will review a draft project pay request, including line items for each subcontractor, with the architect, bank, and/or client on the 25th of the month.

8. The formal/corrected GC payment request will be sent to the architect and then the client for approval by the 30th of the month.

9. The client will pay the GC by the 10th or 30th of month two.

10. The GC's home office accounting department will then release checks (or pay electronically) to the subcontractors ten days after receipt of payment from the owner. Subcontracts often include a 'pay-when-paid' clause which allows the GC to hold off paying subcontractors until the client has paid them. This keeps the GC from having to use its own cash. Contractors all endeavor to operate in the black, which means they have a positive cash flow. Conversely, if the GC operates in the red, it has a negative cash flow and will need to borrow from the bank or other equity partners. See methods to improve cash flow discussed later.

11. Third-tier subcontractors and suppliers are then paid ten or 30 days after the initial subcontractor receives its payment which could be as far off as month three from when materials were delivered, or work performed. Refer back to Figure 11.3.

Jobsite general conditions costs

1. Jobsite general conditions include indirect materials, equipment rental, and indirect labor expenses as was introduced in Chapter 5. The invoicing of indirect materials and equipment is similar to short-form purchase order materials and equipment.

2. Indirect labor is comprised of salaried personnel such as the superintendent, project manager, and jobsite cost accountant. Hourly direct craftsmen also perform indirect cost activities such as cleanup, forklift operation, and other temporary support activities.

3. The accounting for the direct craftsmen performing indirect activities is the same as that for craftsmen performing direct work as discussed earlier.

4. Salaried personnel, including the PM, superintendent, and cost engineer, generally do not complete time sheets as foremen do for their crews. They are individually assigned to a separate cost code within the jobsite general conditions estimate. The home office routinely records a portion of their monthly salary against their individual cost code which is incorporated into the labor report and job cost history report. In some cases, a PM may complete time sheets for his or her jobsite team.

5. Salaried personnel are generally paid twice monthly, and those intervals vary between firms. For example, assume they work through the 15th of month one and then are paid a week later for that half-month, on or about the 22nd. Their salary associated with the second half of month one would then be paid by the home office payroll clerk on or about the 7th of month

two. Often payroll for salaried personnel today is handled with a direct deposit made into the employee's bank account and not with a conventional paper check.

Jobsite revenue

As shown earlier, there are several jobsite expenditures which draw down the contractor's cash flow throughout the month; conversely, there is only one source of positive cash influx and that is receipt of payment from the client. This is usually once monthly, on or about the 10th of the month after the work was performed. In some cases, clients do not pay until the 30th of the month following the month the work was performed. The pay request process is discussed in the next chapter. In the case of speculative residential home builders, they receive one check at the sale or close of each home. Cash flow for speculative builders is therefore very erratic and often requires them to rely on a construction loan to make weekly and monthly payments until a home is sold. Custom home builders will receive monthly draws from the client similar to commercial contractors. Custom home builders and remodeling contractors will sometimes receive a down payment from the bank or the client which allows them some opportunity to operate in the black through the course of construction. So, although 'cash flow' for many contractors follows a typical bell shape or S-curve as presented in Figure 13.1, the revenue curve for a typical commercial GC is a stepped curve in that it is flat and then jumps vertically once monthly as is shown in Figure 13.2.

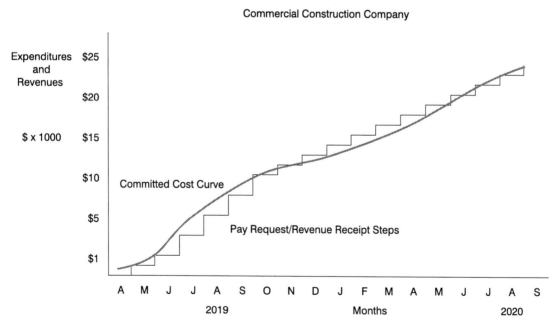

Figure 13.2 Revenue and expenditure curves

Net cash flow and impacts to home office

The CFO processes cash flow as needed to make payments on all of the contractor's jobsite expenditures and receives monthly checks from all of its clients. The contractor's officers and front office rely on a positive and not a negative balance of cash flow. Positive cash flow occurs when revenues exceed expenditures. They expect more money coming in than going out. The CFO will have an established line of credit with their bank and in the case of a short-term cash short-fall they will call on those funds but pay them back as soon as possible. Unless the construction company is owned by an individual and operates as a sole proprietorship, the company officers are typically accountable to a board of directors and equity investors and partners. These equity partners do not expect to have to dig into their pocket when the contractor is operating in the red; instead they expect a positive cash flow and an above-market return on equity due to the high risk of owning and investing in construction companies.

Revenues typically lag expenditures early in the project when there is an increased amount of direct labor expended to erect the structure. The general contractor pays direct labor on a weekly basis and does not withhold retention from the craftsmen's checks. Later in the project, such as during finishes, there is an increased mix of subcontractors from which retention is being withheld and the GC's cash flow needs are eased. The GC's revenue curve will rise above the expense curve when there are more subcontractors than direct labor and material expenditures. Similar to plotting the cash flow curve, there is no exact date this will occur, but it is usually approximately mid-way through construction for most typical commercial construction projects which employ 80–90% subcontractors.

Methods to improve cash flow

As stated repeatedly, general contractors are not in the business of providing construction loans to their clients. Speculative home builders will obtain loans and will pay the bank back when they sell the house – they do operate in the red and the associated interest costs are factored into their home prices. Real estate developers also obtain construction loans and their loan payments are factored into their pro forma as will be discussed in Chapter 19. Commercial GCs expect that their clients will obtain construction loans so that they can keep up with their monthly pay requests; in fact verification that the owner has financing in place is a good contract execution prerequisite. A long list of jobsite expenditures and the processes necessary to approve invoices and make payments was discussed earlier, but there is only one source of revenue which occurs monthly and up to 40 or more days after some contractor costs were expended or committed. So how does a contractor operate in the black and not have to borrow money? A variety of potentials for the GC to improve its cash flow position at the jobsite level, some of them ethical and legal, and are acceptable if negotiated into the contract terms, and others which are very unethical are also offered here. There may be serious ramifications for the GC from the client or even from subcontractors if any of the second bullet-list is used and the contractor's methods were discovered. We are not advocating for any contractor to incorporate them into its project plan.

Ethical cash improvement methods

- Subcontractors finance the general contractor in that subs are not paid until the client pays the GC and retention is held from subs the same as what the client holds from the GC. Employing more subcontractors and reducing the mix of direct labor, direct material, and equipment rental improves a GC's cash flow position.
- Include a mobilization charge in the schedule of values. This is standard with heavy-civil construction contracts.
- Require the client to make a down payment. This is standard with many types of custom residential projects or remodel projects. Many mechanical and electrical equipment manufacturers require down payments as do elevator subcontractors.
- Invoice the client twice monthly, assuming the contract allows this.
- The client pays the GC faster, say on the 10th of month two and not wait until the 30th, or better yet, on the 5th of month two and not the 10th.
- Even though the first month, or partial month, expenditures may be light, the GC should process an early pay request to a) improve its cash flow, and b) test the pay request process as will be discussed in the next chapter.
- Reduce retention held from 10% to 5%.
- The GC does not withhold retention on labor, material, equipment, or indirect costs; therefore, eliminate the client's retention held on these costs. Retention is then only held on the GC equal to the amount of retention held on its subcontractors.
- Release retention on portions of the work which have been completed and accepted. This is common with early subcontractors such as excavation, shoring, and site utilities. This early release requires to have been negotiated into the prime contract prior.
- If the contractor beats their estimate and performs work for less cost, and they are billing against an established schedule of values, they will improve their cash flow position. This is not possible for cost plus percentage fee or time and material projects where invoices are based on actual and not estimated or scheduled costs.
- Some contracts may allow the GC to invoice all of its insurance and other markups up front as a lump sum from the client, and then make distributions periodically.
- Fee will be invoiced proportionately with each monthly draw. The fee will not be paid out but will be banked by the contractor.
- Some GCs negotiate a higher fee to be paid on their direct work than subcontracted work as direct work is riskier, thereby improving its bank of fee as many of these tasks occur earlier.
- Most clients understand that GCs are not banks and will allow a small amount of front-loading of the SOV as discussed in the following, just not an excessive amount.

Unethical cash improvement methods

1. Modify the estimate before developing the schedule of values to reflect more costs on front end construction activities. For example, if the foundations had been originally estimated at $90,000 and the hotel room kitchenettes at $350,000 but reporting those amounts as $300,000 and $140,000 respectively on the SOV.

2. Different than falsifying the SOV as mentioned before, front-loading the SOV also is accomplished by placing a disproportionate share of jobsite general conditions and fee and other markups such as insurance and taxes on those activities which occur early in the project. This seldom can occur on open-book negotiated projects where these markups are invoiced as separate line items as discussed in the next chapter.
3. Increase the amount subcontractors are billing on their monthly requests and falsely report those amounts on the GC's SOV to the client.
4. Hold money back from the subcontractors' checks beyond the retention held due to unresolved disputes or back charges, even though the subcontractors' invoices were applied to the SOV at face value and approved by the project owner.
5. Hold the subcontractors' money long after the GC is paid. Ten days after the client pays the GC is our suggested time-period for a team-build project, but some GCs pay their subcontractors 30 days after they are paid, and some subcontractors do the same with their third-tier subcontractors and suppliers.
6. Hold more retention from subcontractors than the client holds from the GC. For example, if the client holds 5% on the GC then the GC holds 10% from the subcontractors. This is usually not allowed by contract.

Summary

Cash is one of the most powerful tools a client has with a general contractor and a GC has with its subcontractors. The lack of a positive flow of cash is one of the most common reasons for contractor bankruptcy. Contractors which deliver quality work, safely, on time, and within their budget but must continually rely on the bank to fund its operations will ultimately fail. One significant difference between speculative home builders and real estate developers from commercial contractors is that they obtain construction loans, and once the project is completed they realize a large influx of positive cash flow. But the interest they pay on the construction loan was factored into their estimates and pro forma. In commercial construction, as with heavy-civil and custom home construction, contractors expect that the client has either obtained the construction loan or has other sources of funding on hand.

Development of a cash flow curve is a simple task and one that the project manager should do to assist the owner and the bank with analyzing future financing obligations. The cash flow curve is charted from a cost loaded schedule which is developed from the contractor's summary schedule and summary estimate. There are several different cash flow curves which can be plotted, but the one that is most common and straight forward from the contractor's perspective is the work-in-place curve. Processing monthly invoices and receipt of payment will follow a month or so behind this curve.

Net cash flow is defined as revenues less expenditures. Revenue generally occurs once monthly for commercial contractors on the 10th of the month following when the work was performed. Expenses occur at a variety of times throughout the month. Jobsite expenditures include labor, material, equipment, subcontractors, and general conditions. Invoices are processed for all of these expenditures on slightly different tracks involving support from the home office accounting department and the jobsite cost engineer. In most cases, payments lag material delivery and installation by a week or two. In the case of subcontractors, they are paid ten days after the GC is paid by the client. This process is known as 'pay-when-paid' and helps the GC with management of their positive cash flow. There are several methods that the contractor can use to improve its goal of operating in the black with a positive flow of cash. Some of these are ethical and some of these are unethical. The best method to improve cash flow is to prepare timely, accurate, and fair open-book pay requests.

Review questions

1. How can a cash flow curve be used negatively by a client or bank towards a contractor?
2. Why is direct construction work performed with a contractor's own work force considered riskier and therefore they expect a higher fee?
3. Why should PMs view cash flow curves as a 'get-to' and respond gladly when a client requests one be developed?
4. The cost loaded schedule is an intermediate step in developing a cash flow curve. What are two documents necessary to prepare the cost loaded schedule?
5. Should a cash flow curve always be a perfect bell-shape curve? Explain why or why not.
6. There are three different methods for how jobsite general conditions are factored into the cash flow projection and are ultimately invoiced. Name them.
7. How would the cash flow curve presented for Evergreen Construction Company change if:
    a. The project start date was delayed for one month,
    b. All of the site improvements were re-scheduled to be accomplished during early shell construction rather than near the completion of the project,
    c. The swimming pool design was finalized and will cost $350,000 which is change ordered into the contract and will be installed during the last month of the schedule, or
    d. Retention will only be held on Evergreen's subcontractors?

Exercises

1. What would happen to a PM who a) operated in red for the first time on his or her construction project, or b) repeatedly operated in the red on all of his or her projects?
2. What would happen to a PM who was $1 million overbilled on a) a lump sum project, and/or b) a negotiated guaranteed maximum price project with open-book accounting and was discovered by the client's auditor during a monthly audit? You may want to peek ahead at the chapters on pay requests and audits for this question.
3. Is a GC-subcontractor 'pay-when-paid' contract clause ethical? Is it legal?
4. What would happen to a GC's cash flow if it had been scheduled to receive its invoice from the client on the 30th of month two but did not receive it until the 30th of month three? What would happen to subcontractors of subcontractors if all were subsequently paid 30 days after receipt of payment?
5. A complete contract with special conditions and general conditions was not included in this text due to space limitations. Perform some outside research and cite reasons and specific article numbers in the A102 contract and A201 general conditions which protect the contractor in Exercise Four.

6. List one or more ethical and legal possibilities a contractor could use to improve its management of jobsite cash flow beyond those discussed here.

7. Prepare an argument for your client why you should receive a 10% down payment before you start any work on their project.

8. Assume a GC has the following expenses for month one of their project. Will they be in the red or black when they receive payment, less 10% retention but including a 5% fee, on the 10th of month two and by how much?

- Direct labor of $40,000 committed to evenly throughout the month;
- Short-form PO direct material received and invoiced on the 15th for $20,000;
- Long-form PO materials are received and invoiced on the 15th for $40,000;
- $100,000 worth of subcontracted work performed and invoiced on the 20th which include a matching 10% retention clause. Subcontractors are paid ten days after the GC is paid;
- Indirect labor at $30,000 per month payable half on the 22nd of month one and the second half on the 7th of month two; and
- Indirect material costing $10,000 and invoiced on the 30th of the month.

9. Assume the same expenditures as Exercise Eight but now the client does not pay until the 30th of month two. Is the GC now in the red or black and by how much?

10. Assume the same expenditures as Exercise Eight but now the GC holds the subcontractor and major supplier checks for 30 days after the GC has been paid by the client on the 10th. Is the GC now in the red or black and by how much?

11. Draw a cash flow chart for a fourth-tier supplier such as the mechanical subcontractor's ductwork insulation subcontractor's material supplier. You may need to draw a quick organization chart. Assume the ductwork insulation is installed on May 1. Assume standard contracts and pay periods and processes and predict a) when is the earliest the supplier might be paid, and b) when is the latest the supplier might be paid?

14

Payment requests

Introduction

Construction project managers are responsible for many jobsite operations, but getting paid for the work performed is one of the most important. A project manager (PM) may have all of the tools necessary to earn a profit on a job, but if the owner does not pay for the work, the contractor will not be able to realize a profit. Some PMs do not acknowledge the importance of preparing prompt payment requests. This is especially true with many subcontractors. If a payment request is not submitted on time, the contractor will likely not get paid on time. Cash management is essential, or the general contractor (GC) may find that they are unable to pay suppliers, crafts-men, or subcontractors. The importance of a positive cash flow was the topic of the last chapter. PMs must be able to manage jobsite cash flow to be effective contributors to the operation of the construction company.

Processing pay requests is one of the most important aspects of construction financial manage-ment for the project manager. Although presented as a jobsite activity in this book, the PM will receive support from the home office accounting department. There are many aspects of construc-tion management that relate to and are involved with the pay request process including contracts, schedule of values, retention, and lien management. This chapter will discuss all of these elements and others affecting the contractor's progress payment requests and associated interaction with the cost accounting processes. Some members of the built environment industry may use different terms for this important topic such as pay estimates, invoices, bills, draws, and progress payments; they are all considered similar for this discussion. Much of the material from this chapter has relied upon J. Schaufelberger and L. Holm's *Management of Construction Projects, a Constructor's Perspective (MCOP)* (2017). The reader is suggested to review that resource for additional discussions on payment requests and other more advanced project management processes.

Contract types

As stated previously, the construction industry differs from others for several reasons, and one of those is the monthly payment process. The formats and times for the pay request are specified in the supplemental or special conditions of the contract. Payment procedures used on the Olympic Resort and Hotel are contained in Article 12 of the AIA A102 cost plus fee with a guaranteed maximum price (GMP) contract. Regardless of the type of contract, many of the payment procedures are similar. This section will discuss some of the pay request differences that are affected by different types of contracts.

Open-book *cost-plus guaranteed maximum price construction contracts* differ from lump sum contracts as regards to the monthly pay request process. The project manager requests payments based on actual and projected expenses. He or she must have already received invoices from subcontractors and suppliers. The PM generally is required to submit subcontractor invoices and general contractor payrolls to the project owner as backup with the payment request. Fees, direct work items, and general conditions expenses are all billed using a schedule of values (SOV) and percentage of completion. It is almost impossible to overbill open-book projects as they are often subject to periodic owner audits of actual costs incurred.

Payment on a *lump sum contract* is also based on a pre-established schedule of values and percentage completion. Front-loading and overbilling can occur on this type of contract, as will be discussed later. General contractor records are rarely audited in a closed-book lump sum contract. A unit price contract allows for payment based upon quantities actually installed. If the contractor is to be paid $200 per lineal foot (LF) for water line installation in a remote location, and has installed 100 feet, then they will be paid $20,000 (100 LF x $200/LF) less any agreed upon retention. This process is quite objective and can be facilitated by an outside quantity measurement individual or firm. Payment on a time and materials (T&M) contract is based on actual labor hours multiplied by a labor rate plus reimbursement for materials based on supplier invoices. A loaded wage rate on a T&M project includes labor burden plus markups for overhead and profit.

Schedule of values

The first step to develop a construction pay request is to establish an agreed upon breakdown of the contract cost, or schedule of values. Often the contract will require that a SOV be submitted for approval within a certain time after executing the contract, for example within one week. This SOV should be established and agreed upon early in the job, well before the first significant request for payment is submitted, but only after all subcontracts have been awarded. If the SOV is established before subcontractor buyout it will not necessarily be accurate and may be difficult to invoice against.

One common method to develop the SOV is to first start with the summary estimate, similar to the process to prepare the cost loaded schedule. This is shown as the GMP cost column in the center of SOV worksheet Table 14.1. This would be the SOV used on a cost-plus contract because the general conditions and fee are listed separately. On a lump sum contract, the general conditions and fee would be distributed proportionately across all payment items as shown on the right side of Table 14.1. The SOV that the project manager would submit for a lump sum contract is the

Table 14.1 Schedule of values worksheet

	Evergreen Construction Company				
	SOV Worksheet				
	Olympic Hotel and Resort, Project 422				
CSI Division	Description	GMP Value	% of Subtotal	GC & Fee Pro-rated	Adjusted Totals
	Reinforcement & PT	2,104,399			
	Foundations	91,177			
	Walls and Slabs	2,631,036			
3	Concrete Subtotal:	4,826,612	24.3%	1,124,992	**5,951,604**
4	Masonry	150,000	0.8%	34,962	**184,962**
5	Structural and Misc. Metals	411,103	2.1%	95,820	**506,923**
	Rough Carpentry	1,505,280			
	Finish Carpentry	1,613,913			
6	Carpentry Subtotal:	3,119,193	15.7%	727,025	**3,846,218**
	Insulation	182,257			
	Roof and Accessories	573,638			
	Waterproofing	484,333			
	Sheetmetal	154,586			
7	Total Thermal and Moisture:	1,394,814	7.0%	325,105	**1,719,919**
	Doors and Frames	390,720			
	Glazing	720,920			
	Door Hardware	173,300			
8	Division 8 Subtotal:	1,284,940	6.5%	299,495	**1,584,435**
	Drywall and Lobbies	1,283,000			
	Painting	390,723			
	Floor Covering	663,343			
9	Finishes Subtotal:	2,337,066	11.7%	544,726	**2,881,792**
10	Specialties	157,085	0.8%	36,614	**193,699**
11–13	Equipment	355,237	1.8%	82,799	**438,036**
14	Elevators	435,000	2.2%	101,390	**536,390**
21	Fire Protection	442,700			
22	Plumbing	1,211,379			
23	HVAC and Controls	720,629			
21–23	Mechanical Subtotal:	2,374,708	11.9%	553,499	**2,928,207**
26	Line Voltage Electrical	1,286,702			
27	Communication and Security	332,500			
26, 27	Electrical Subtotal:	1,619,202	8.1%	377,405	**1,996,607**

Evergreen Construction Company

SOV Worksheet

Olympic Hotel and Resort, Project 422

CSI Division	Description	GMP Value	% of Subtotal	GC & Fee Pro-rated	Adjusted Totals
31, 33	Earthwork and Utilities	906,771			
32	Paving	205,550			
32	Misc. Site and Lscape	315,990			
31–33	Site Work Subtotal:	1,428,311	7.2%	332,912	**1,761,223**
	Subtotal w/o GC's and Fee:	19,893,271	**23.3%**	**4,636,744**	**24,530,015**
	Labor Burden	1,074,287			
1	Jobsite General Conditions	1,662,783			
	Fee & Insurance & Excise & Cont.	1,899,674			
	Subtotal Markups	4,636,744			
	TOTAL GMP:	**24,530,015**			

far right column titled 'adjusted totals'; none of the other columns would be shown. The same cost codes which match those from the estimate and the schedule should be applied here as well.

By combining line items within the SOV together, with the result being fewer line items, some contractors believe they can overbill or hide true estimates or cost values from the project owner. The SOV should be as detailed as is reasonable. The project manager should do all that is possible to assist the owner in paying completely and promptly. Nothing should be hidden. At a minimum, the former 16 Construction Specifications Institute (CSI) divisions, or relevant divisions from the new 49 CSI divisions, should be used as line items. Major subcontractors should be listed where possible. Separate building components, building wings, distinct site areas, phases, or systems should be listed individually in a detailed SOV. A narrow view or summarized SOV for a closed-book project might look like Table 14.2. An abbreviated or summary SOV like this may make it difficult for the PM to sell the monthly pay estimate to the owner and the bank. An open and honest pay request process facilitates the contractor's ultimate goal of timely payments.

It is a good practice for the project manager to submit the proposed SOV to the project owner even if the contract does not require it. The PM does not want any future arguments with an owner or architect over a payment request. The SOV should be submitted for approval, just as a door hardware schedule would. Most project owners appreciate a contractor's transparency which will help facilitate prompt payment as well as establish necessary respect and trust.

The fee on cost-plus guaranteed maximum price contracts is usually invoiced as a percentage complete that matches the overall project completion level. If the project is 80% complete, then 80% of the fee has been earned. Most owners will not take issue with this approach. General conditions on cost-plus contracts can be invoiced three different ways as was discussed with the cost loaded schedule preparation in the last chapter, including: straight line with equal payments for each month, percentage complete based upon work constructed, or actual costs incurred.

Table 14.2 Summary schedule of values

Evergreen Construction Company
1449 Columbia Avenue, Seattle, WA 98202
(206) 447-4222
Olympic Hotel and Resort, Project 422
4/1/2019
Summary Schedule of Values

CSI Division	Description	Totals
3	Concrete	$5,951,604
4	Masonry	$184,962
5	Structural and Misc. Metals	$506,923
6	Carpentry	$3,846,218
7	Thermal and Moisture	$1,719,919
8	Doors and Glass	$1,584,435
9	Finishes	$2,881,792
10	Specialties	$193,699
11–13	Equipment and Furnishings	$438,036
14	Elevator	$536,390
21–23	Mechanical Systems	$2,928,207
26, 27	Electrical Systems	$1,996,607
31–33	Site Work	$1,761,223
	Total Bid:	**$24,530,015**

Once change orders are approved their values can either be spread across the applicable schedule of values line items or added to the bottom as new line items. This second method generally is the easiest to administer, but this complicates the owner's ability to track subcontractor monthly and final lien releases. Conversely, reformatting the SOV each month by spreading change orders may cause the record of the original SOV to be lost.

Some contractors, especially those which specialize in lump sum bidding, may advocate hiding the fee and general conditions, or front-loading them. This is more prevalent on a bid contract than with negotiated work. We are recommending that each line item, including the fee and general conditions, be listed just as they would in the project cost accounting system. The schedule of values should look like the contractor's estimate. Trying to explain during an audit or a claim situation why the cost of the under-slab utility work was stated as $50,000 in the pay estimate, but was only $20,000 in the original bid, will be difficult if the contractor is found out. Spreading, but still hiding, the fee and general conditions as a weighted average over the SOV is common for both guaranteed maximum price and lump sum projects, and although this may be fair, it will be a difficult process for the project owner to track lien releases as discussed later.

Payment request process

Construction invoicing usually occurs at the end of each month. The jobsite cost engineer will assist the project manager and gather all of the costs in preparation of the monthly request that is to be submitted to the architect or owner for approval. This process should start on or about the 20th of the month. Subcontractors and major suppliers should be required to have their monthly invoices turned in to the general contractor by that time. Subcontractors often do not do a good job of managing their cash flow. The PM needs to encourage them to submit their monthly billings timely. Some GCs have the attitude that if the subcontractors do not get their invoices in on time, that is the subcontractors' problem and they will not get paid this month. While this may be contractually correct, it is counter-productive. The contractor's jobsite team must do all that is reasonable to keep the subcontractors with enough cash so that they do not suffer bankruptcy, at least while they are on this project. A senior executive at a major GC indicated his firm personally calls their subcontractors each month on the telephone – not by email – to remind them to get their invoices turned in on time.

The *subcontractors' invoices* received on the 20th of the month should reflect what percentage of work they believe they will have completed and in place through the end of the month. The suppliers also estimate what they plan to deliver to the site by the end of the month. These timelines will have been established in their contract agreements. The general contractor's project manager and superintendent, with the assistance of a jobsite cost engineer, also forecast the estimated cost of direct work activities they plan to have in place through the end of the month.

Subcontractors and suppliers often request payment for materials which have been delivered to their own warehouse but are not yet on the jobsite. Maybe fabrication is necessary, as is the case with hotel room kitchenettes or roof trusses, or maybe the supplier is ahead of schedule or the general contractor is behind. Sometimes payment for *off-site stored materials* is unavoidable due to scheduling reasons. Occasionally it may be financially beneficial for all parties. For example, the mechanical subcontractor was able to purchase the kitchen exhaust stainless steel ductwork at a discount because it was purchased with materials needed for another larger project. But payment for materials stored off the project site has complications associated with potential damage or theft. In these cases, the project manager needs to be sure that his or her interests, and that of the owner, are protected. The material must be stored in an insured and bonded warehouse. This also requires a personal inspection and verification. The jobsite cost accountant could help with this as well. Most GCs, architects, owners, and lenders try to avoid paying for off-site stored materials but it can be in all of their best interests if accounted for properly.

On approximately the 25th of the month the project manager and cost engineer collect all of these cost forecasts, estimates, and payment requests and assemble a *draft pay request* to submit to the architect, project owner, and/or the bank for review. This involves estimating the percentage complete for each item on the schedule of values through the end of the month. These percentages are multiplied by the value of each line item to produce the SOV continuation sheet shown in Table 14.3. The format shown is an Excel spreadsheet customized by Evergreen Construction Company for the Olympic Hotel and Resort project. Many contractors use similar spreadsheets for their monthly SOV submissions. A live version of this spreadsheet is available on the eResource. Alternatively AIA Form G703 or ConsensusDocs® Form 239 may be specified in the contract as the required SOV worksheet.

Table 14.3 Pay request schedule of values continuation sheet

Evergreen Construction Company	Billing Period: From:	11/01/19
1449 Columbia Avenue, Seattle, WA 98202	To:	11/30/19
(206) 447-4222	Request #	8
Olympic Hotel and Resort, Project # 422	Invoice #	2927

PAY REQUEST SOV CONTINUATION SHEET

A	B	C	D	E	G (D+E)	G.1 (G/C)	H (C–G)
Line Item	Description	Contract Value	Previous Complete	Complete This Mo.	Complete To date	% Complete	To Go
1	Jobsite General Conditions	1,662,783	831391	103924	935,315	56%	727,468
3.1	Reinforcement & PT	2,104,399	2,104,399	0	2,104,399	100%	0
3.2	Foundations	91,177	91,177	0	91,177	100%	0
3.3	Walls & Slabs	2,631,036	2,631,036	0	2,631,036	100%	0
4	Masonry	150,000	0	0	0	0%	150,000
5	Structural & Misc. Stl.	411,103	411,103	0	411,103	100%	0
6.1	Rough Carpentry	2,223,633	1,404,000	404,000	1,808,000	81%	415,633
6.2	Finish Carpentry	895,560	0	0	0	0%	895,560
7.1	Insulation	182,257	0	0	0	0%	182,257
7.2	Roof & Accessories	573,638	0	66,000	66,000	12%	507,638
7.3	Waterproofing	484,333	0	33,000	33,000	7%	451,333
7.4	Sheetmetal	154,586	0	0	0	0%	154,586
8.1	Doors & Frames	390,720	0	50,000	50,000	13%	340,720
8.2	Glazing	720,920	0	67,000	67,000	9%	653,920
8.3	Door Hardware	173,300	0	25,000	25,000	14%	148,300
9.1	Drywall	1,283,000	0	0	0	0%	1,283,000
9.2	Painting	390,723	0	0	0	0%	390,723
9.3	Floor Covering	663,343	0	0	0	0%	663,343
10	Specialties	157,085	0	0	0	0%	157,085
11	Equipment	355,237	0	0	0	0%	355,237
14	Elevators	435,000	100,000	0	100,000	23%	335,000
21	Fire Protection	442,700	0	67,000	67,000	15%	375,700
22	Plumbing	1,211,379	0	268,000	268,000	22%	943,379
23	HVAC & Controls	720,629	0	55,000	55,000	8%	665,629
26.1	Line Voltage Electrical	1,286,702	0	534,000	534,000	42%	752,702
26.2	Low Voltage Elect	332,500	0	0	0	0%	332,500
31	Earthwork & Utilities	906,771	560,000	0	560,000	62%	346,771
32.1	Paving	205,550	0	0	0	0%	205,550

	Evergreen Construction Company				Billing Period: From:	11/01/19

Evergreen Construction Company
1449 Columbia Avenue, Seattle, WA 98202
(206) 447-4222
Olympic Hotel and Resort, Project # 422

Billing Period: From: 11/01/19
To: 11/30/19
Request # 8
Invoice # 2927

PAY REQUEST SOV CONTINUATION SHEET

A	B	C	D	E	G (D+E)	G.1 (G/C)	H (C–G)
Line Item	Description	Contract Value	Previous Complete	Complete This Mo.	Complete To date	% Complete	To Go
32.2	Misc. Site & Lscape	315,990	0	0	0	0%	315,990
	Subtotal Cost:	21,556,054	8,133,106	1,672,924	9,806,030	45%	11,750,024
90.0	Fee and other Markups	2,973,961	1,122,369	230,864	1,353,232	45%	1,620,729
99.1	Change Order No. 1	75,000	0	0	0	0%	75,000
99.2	Change Order No. 2	117,295	27,550	44,567	72,117	0%	45,178
	Current Totals	**24,722,310**	**9,283,025**	**1,948,355**	**11,231,379**	**45%**	**13,490,931**

Once the draft pay request is assembled, the project manager and jobsite cost engineer should schedule a short informal meeting with the architect, the owner, and the bank at the jobsite to review this month's proposed invoice. Often each of these *draft monthly pay request meetings* are scheduled for the entire project at the time of the initial project preconstruction meeting. At each of these meetings the pay request is presented as a draft for discussion and approval. If any of the approving parties has a problem with a particular line item or percentage, the general contractor still has time to request an explanation from a subcontractor or develop additional detail. A job walk during this meeting is extremely helpful to visualize the work completed or that will be in place by the end of the month. If necessary, subcontractor invoices can be attached to this draft for backup. This draft pay estimate and the meeting promote teamwork among the owner, the architect, and the GC's jobsite financial team. Similar pay request review processes are recommended for both bid and negotiated projects to facilitate timely payment.

The general contractor's project manager will submit the *formal pay request* for final approval and payment to the owner no later than the end of the month. This can be submitted earlier if possible but this date will have been prescribed in the prime contract agreement. ConsensusDocs® Form 291 and AIA Form G702 are popular copyrighted pay estimate summary coversheets; either of which may be stipulated as required in the contract special conditions. Figure 14.1 was developed by the hotel case study client as a customized example of a pay request summary and approval page which accompanies the detailed schedule of values shown in Table 14.3. This client builds hotels all across the country and has many standardized forms and procedures for managing its construction projects. Some contracts and lenders may also request that the architect sign off in the approval process. Although electronic submissions are popular today, it is still a good practice for the PM to hand carry the final payment request through the approval process to ensure there are not any issues.

PAY REQUEST SUMMARY

Where the basis of payment is a guaranteed maximum price

Client: Northwest Resorts, LLC
Address: 1001 First Ave., Seattle, WA 98202
(206) 392-5336

Contractor: Evergreen Construction Company
Address: 1449 Columbia Ave., Seattle, WA 98202
(206) 239-1422

Project:	Olympic Hotel and Resort	Contractor's Project #	422
Billing Period: From: 11/01/19		Contractor's Request #	8
To: 11/30/19		Contractor's Invoice #	2927

Original Contract Price:		$24,530,015
Approved Change Orders: 1 through 2		$192,295
Current Contract Price:		$24,722,310
Total work completed to date: (See attached detailed SOV)		$11,231,379
Less total retention held at a rate of:	5%	−$561,569
Plus applicable State Sales tax added at a rate of:	10.00%	$1,123,138
Subtotal completed to-date less retention, inclusive of Sales Tax:		$11,792,948
Less prior applications for payment, net of retention and including tax:		$9,747,176
Net amount due this pay period:		$2,045,772

Contractor's Certification:
I hereby do certify that to the best of my knowledge the above accounts reflect a true and accurate estimate of the values of work completed and that all amounts paid prior have been distributed according to existing contractual requirements and the laws in existence at this jurisdication at the time of contract.

Contractor: *Chris Anderson*   Date: 11/30/19

Architect's Certification:
In accordance with the contract documents and limited on-site evaluations we have no reason to believe that the accounts presented by the contractor are not reflective of the work accomplished to date.

Architect: *Stan Zimmer*   Date: 12/01/19

Figure 14.1 Pay request summary

The prime contract will also describe *payment terms*. It is customary for the general contractor to be paid by the 10th of the following month if the payment request is submitted by the end of the current month, but some project owners pay on the following 30th, which puts a cash flow strain on the construction team. The specific time frames used on the Olympic Hotel are contained in Article 12 of the AIA A102 contract. The GC's project manager should volunteer to pick up the monthly check and hand-deliver it to his or her CFO and avoid the 'check's in the mail' scenario. Today many contractors will incorporate a direct electronic deposit process in their contract to expedite this step. After the GC has received the monthly payment from the owner, it should disburse funds to suppliers and subcontractors expeditiously. This usually is also done within ten days of receipt. The GC should not delay paying suppliers and subcontractors, but endeavor to keep them financially solvent.

Some third-tier subcontractors and suppliers will request *joint check agreements* between themselves and the second-tier subcontractor and the general contractor to ensure that they are paid expeditiously. In this case, one joint check is cut naming the GC and both the primary subcontractor and its supplier or third-tier subcontractor as recipients. Although some GCs see this as an arduous task, joint checks ultimately benefit and protect the GC and the project owner from the potential that second-tier subcontractors may not make payments to third-tier firms which may ultimately lien the project.

Cash as a tool

As discussed in detail in the last chapter, cash is one of the most powerful tools the project owner has with the general contractor and the GC has with their subcontractors. If a construction firm has a reputation of paying their subcontractors and suppliers promptly, they may receive more favorable prices on bid day. This will result in the GC getting more work. GCs (the same applies here for subcontractors) are not banks. The GC's role on the construction project is to build the building, not provide construction financing. Each contractor begins incurring labor and material expenses early in the month. Bills for these expenses are submitted at the end of the month, and payment is not received until the 10th or the 30th of the following month. This means the GC has provided funding for these expenses for a minimum of 40 and maybe up to 60 days. These time periods may be up to 30 days longer for subcontractors.

Unfortunately slow or delayed payments from project owners to GCs and from GCs to subcontractors motivate all contractors to attempt to *front-load and overbill*. The consequence of overbilling is that contractors may be caught and trust and faith in the jobsite team are lost by the project owner and the bank. The owner may allow a little overbilling so as not to place the contractor in a situation with too large a negative cash flow. The project manager should cooperate by not overbilling too much, as if discovered it might impede the monthly payment request process. The way some PMs overbill is to front-load the schedule of values. Ways of accomplishing this are to place all of the fee and general conditions (if not separately listed) on the early scheduled construction activities, such as foundations or concrete slabs. Another way is to artificially inflate these early activities or to falsify the amounts requested by subcontractors. These methods are counter-productive to building with a team attitude and will most often ultimately be discovered through interim lien releases or during an audit. An exercise has been included at the end of this chapter for front-loading a SOV for the interested reader, but the practice is not advocated by this author/contractor.

The use of *discounts* in the pay request process is another example of using cash as a tool. Discounts may be offered by material suppliers for early payment. If the general contractor pays within a certain time period, say ten days after receipt of an invoice, they can receive a small discount off the invoice price. For this reason, a project manager may elect to pay suppliers early, even though the owner has not made a payment. In this case, the GC is acting as a bank. Some owners will want to benefit also from these discounts. Discounts should only pass through from suppliers to the project owner if the owner has paid the GC up front, and the GC is not utilizing its own cash to realize the discount.

Retention

A portion of each monthly pay request is customarily held back by the project owner from the general contractor and subsequently held back by the GC from its subcontractors. This is referred to as retention, or retainage, and is another cash tool. The purpose of retention is to assure that the contractor's attention is focused on completion of the project, including close-out of all physical and paperwork activities. Financial close-out and the process utilized to receive release of final retention are discussed in Chapter 16.

The percentage of retention held will be dictated by the prime contract and on private projects may be a subject for negotiation, or, during the course of the project, re-negotiation. In the past, 10% was standard, but as the size of contracts has increased along with the value of money, 10% is considered by many as excessive in today's market. The most common percentages today are either 5% through the course of the job or alternatively 10% until the project is 50% complete and no further retention held thereafter, which also equals 5% upon completion. Paragraph 12.1.7 of the AIA A102 contract agreement for the Olympic Hotel indicates a retention rate of 5%.

Each subcontractor's contract agreement with the general contractor should be tied into the prime contract agreement between the GC and the project owner. In this way, the subcontractor is usually tied into whatever retention terms the GC has with the owner. This is usually in the subcontractor's best interest and prohibits the GC from over-withholding retention, which was one of the unethical cash flow improvement options discussed in the last chapter.

If the owner retains 5% of the entire job until 30 days after substantial completion (the standard time frame), does this mean that the foundation shoring subcontractor has to wait to receive final payment until all of the interior painting punch list items are complete? This is not fair to the early subcontractors and may result in undesirable financial burdens that can ultimately impact both the general contractor and the project owner. It is a good idea to include a clause in the prime contract that allows for release of a portion of the overall retention to facilitate paying off early subcontractors. The garage shoring contractor can go broke on another project, but the project manager may receive a lien on this project because the retention is still outstanding. This is not as applicable on short duration projects, but is extremely important on jobs that last a year or longer. Many GCs who pride themselves on team-build projects feel that if final and unconditional lien releases are received 30 days after early subcontractor work packages are completed, those subcontractors should be paid off. But not all project owners subscribe to this philosophy.

Retention establishes an account to finish the work in case contractors refuse or cannot do so. Retention also serves as an incentive for contractors to finish the project expeditiously. A general contractor may be able to convince a negotiated project owner that they do not need to hold as large a retention percentage, because the teams have worked together previously. Coincidently 5% retention approximately equals a mid-size commercial GC's fee. Obtaining the final retention release should be an incentive for any contractor to have a smooth and expeditious close-out process.

Lien management

Liens are legal rights by contractors and suppliers to secure payment for materials and work performed provided in the improvement of real estate. These rights attach to the property in a manner similar to a mortgage. If a contractor is owed money, and the owner refuses payment, the contractor can attach a lien to the owner's property. If the lien is not removed by payment, the lien claimant can demand legal foreclosure and have the obligation satisfied out of the proceeds of the sale of the property. Liens also make it difficult for a property owner to transition from a construction loan to a less expensive permanent loan. Liens should be avoided at all costs. They are expensive to deal with, they cause bad feelings, and they damage reputations. The project manager and cost engineer should work to protect the owner from liens from their own subcontractors and suppliers and from third-tier subcontractors and suppliers. This section describes several lien management or lien prevention techniques which may be implemented by the jobsite financial management team.

Material supplier's rights are protected by state law to assure project owners pay for goods they received. In most jurisdictions, suppliers are required to file what is known as a *materialmen's notice*. This notice is sent to the property owner notifying them that the supplier will be delivering material to their project. In many states the notice is technically required in order to preserve the supplier's lien rights. Some notices are also sent to the general contractor. The cost engineer should file all material notices received with each subcontract and purchase order. At the end of the project, it is appropriate for the GC to request unconditional lien releases from each supplier who filed material notices.

Project owners should require *conditional lien releases* to accompany each request for payment from the general contractor and the GC should require the same from its suppliers and subcontractors. The contract may require submission of interim (or conditional) lien releases from suppliers and subcontractors with each GC payment request to the owner. A project manager should require suppliers and subcontractors to submit conditional lien releases with their monthly pay requests, regardless of the owner's requirements. Interim lien releases should be collected and filed just in case a problem occurs. Regardless of whether a project has a lump sum or a guaranteed maximum price contract, if the project owner requests copies of subcontractor releases, the GC should provide them. The more protection the team can have from liens the better off they are.

A conditional lien release is conditioned upon receipt of payment. If the contractor receives payment for the requested amount, they release their rights to claim for that specific amount. An *unconditional lien release* means that all payments have been received to date, and those lien rights are unconditionally released. Some payment request procedures will require conditional releases to accompany each request for payment along with an unconditional release for the previous month's payment. Some owners will go so far as to physically trade checks for lien releases.

Lien releases can take on a variety of forms and many states produce standard lien release templates which comply with state law. Some owners will try to get contractors to release lien rights that are not required to be released by law. Releasing rights to claim for extra work are also sometimes hidden within the lien release or payment request language. An example lien release is shown in Figure 14.2. Lien releases are an area where legal counsel should be sought. The conditional release covers the month for which payment is requested, and the unconditional release is cumulative for the payments that have been received. A *final lien release* will be submitted for the entire general contractor's contract amount upon close-out to receive the last release of retention.

INTERIM LIEN/CLAIM WAIVER

From: *Evergreen Construction Company*   Project: *Olympic Hotel*
        1449 Columbia Ave.   *Route 4 Box 787*
        Seattle, WA 98202   *Rainforest, WA 98088*

Project Manager: *Chris Anderson*   Contact Person: *Sandi Chambers*
Telephone: *(206) 447-4222*   Telephone: *(206) 392-5336*

CONDITIONAL RELEASE

The undersigned does hereby acknowledge that upon receipt by the undersigned of a check from *Northwest Resorts, LLC* in the sum of *$2,045,722.00* and when the check has been properly endorsed and has been paid by the bank upon which it was drawn, this document shall become effective to release any and all claims and rights of lien which the undersigned has on the above references project for labor, services, equipment, materials furnished and/or any claims through *November 30, 2019*, except it does not cover any retention or items furnished thereafter. Before any recipient of these documents relies on it, said party should verify evidence of payment to the undersigned.

UNCONDITIONAL RELEASE

The undersigned does hereby acknowledge that the undersigned has been paid and has received progress payments in the sum of *$9,747,176.00* for labor, services, equipment, or materials furnished to the above referenced project and does hereby release any and all claims and rights of lien which the undersigned has on the above referenced project. The release covers all payment for labor, services, equipment and materials furnished, and/or claims to the above referenced project through *October 31, 2019* only and does not cover any retention or items furnished after that date.

I CERTIFY UNDER PENALTY OF PERJURY UNDER THE LAWS OF THE STATE OF WASHINGTON THAT THE ABOVE IS A TRUE AND CORRECT STATEMENT.

Signature: *Robert Benson*
                (Authorized Corporate Officer)
                Officer-in-Charge

Date: *November 30, 2019*

Figure 14.2 Lien release

Other techniques the estimating, buyout, and cost engineering team should utilize to *manage the lien process* and protect their client from potential lines include the following:

- Request bids from only prequalified subcontractors and suppliers.
- Utilize efficient subcontractor buyout practices and buy smart, and not simply buy fast by mailing contract agreements to those with the (apparent) lowest quotes from bid day. Make sure each chosen subcontractor is a best-value subcontractor.
- Check subcontractor financial statements; similar to the process a client would practice with a general contractor.
- Draft tight subcontract agreements.
- The most common liens are filed from third- or even fourth-tier companies. These may be from an out-of-state supplier to another supplier to the third-tier subcontractor to the second-tier subcontractor. Complicated? Yes! The farther removed, the more difficult it is to prevent and ultimately resolve. Identify as many third- and fourth-tier subcontractors as is

possible, along with their contract amounts. If two-party checks are necessary, accommodate them.
- Pay subcontractors promptly after being paid by the client.
- Timely negotiate all change orders and back charges.
- Support retention release for early subcontractors.

So what happens if a supplier does not get paid within 30 days of the due date? Usually a *pre-lien notice* is filed which is not a lien, but is just a warning that if the supplier does not get paid by a certain date, it will file a lien. If an additional 30 days passes without payment (now maybe 90 days total since the invoice was sent or materials delivered) the supplier will file a formal lien. After the lien is filed, the supplier has an additional 90 days to foreclose. Many states have different rules with respect to liens and lien timing and foreclosures.

Liens are filed by contractors and suppliers against the owner's property. This appears to be an owner's problem, so why should the general contractor's project manager worry about them? Liens are seldom filed by the GC. It is the owner who has received a lien from the contractor's supplier. The GC is expected to protect the owner from liens. Once a lien is filed it becomes extremely difficult for the GC to process any further pay requests. But, contractors' and suppliers' rights to lien are sometimes necessary construction management tools and the PM and his or her jobsite cost accountant or cost engineer must understand and manage them and associated legal and contractual rules, language, and timing correctly.

Summary

Receipt of timely payment is one of the most important cash flow responsibilities of the general contractor's jobsite financial team. The exact format for submitting payment requests will vary depending on the type of contract. A schedule of values is used to support payment applications for lump sum and cost-plus contracts. The project manager is responsible for developing the payment request, making sure payment is received, and subsequently seeing that the subcontractors and suppliers are paid. The jobsite cost accountant will assist the PM with this. If payment has not been received on time, the PM should contact the owner to determine the cause. The financial relationship with the owner is the GC's PM's responsibility. The same scenario holds true with respect to subcontractors and suppliers; the GC's PM ensures that they are paid promptly. Owners usually retain a portion of each payment to ensure timely completion of the project; the retention rate is specified in the contract. Liens can be placed on a project if subcontractors or suppliers are not paid for their labor or materials; therefore project owners require lien releases to be submitted with monthly payment applications.

Review questions

1. Why should all team members wait until completion of subcontractor buyout to finalize the SOV?
2. Is Evergreen's GMP SOV as presented in the left hand side of Table 14.1 based upon the old or new CSI divisions?

3. What type of data is required to be submitted to support a payment request on a lump sum contract?
4. Why are interim lien releases requested by the GC from subcontractors?
5. Why are subcontractor lien releases requested by the client from the GC?
6. Why might the GC's PM want to pick up the owner's check in person?
7. What would happen to a subcontractor's lien release if the GC unilaterally revised their invoice and processed a check?
8. What would happen if Northwest Resorts discovered Evergreen Construction Company was overbilled by $2 million during a mid-project audit of our open-book case study project?

Exercises

1. Prepare a SOV similar to the one shown in Table 14.3, except for another sample case study project, such as a retail tenant improvement project, pre-cast concrete bridge, or athletic field complex.
2. Prepare a linear bar chart schedule, similar to a construction schedule, for two to three months just focusing on all of the activities and dates associated with the pay request process. You may need to refer back to Chapter 13 for additional dates and activities.
3. Prepare a pay request for the month ending March 2020 (or any other chosen month) for the Olympic Hotel, including all supporting documentation and lien releases. Assume the project is on schedule. Refer back to the cost loaded schedule from Chapter 13. Make whatever assumptions are necessary.
4. We do not advocate this, but for comparison, prepare a lump sum SOV for the Olympic Hotel example except aggressively front-load the general conditions and fee and other markups. How would this change the cash flow curve presented previously in Chapter 13?
5. What are some materials or fabrications that a GC may be willing to pay a supplier for early if the materials were either stored on the jobsite or in the supplier's yard and why those materials?
6. Assume this is a $20 million project and $10 million has been invoiced and received prior to this month. The GC is processing a $2 million invoice for this month. How much will the unconditional lien release be and how much will the conditional lien release be?
7. What should the GC's PM do if one subcontractor does not send their invoice in on time? How would your answer differ if this were a large or a small subcontractor?
8. What should the GC's PM do if the owner has a problem with one subcontractor's proposed invoice during the 25th of the month walkthrough, but that subcontractor is refusing to modify its proposed invoice amount?
9. Proper pay request processing is an important project management and cost engineering responsibility. What are some other important responsibilities for the PM?
10. Assume this is a $10 million construction project and $8 million, including fees and other markups, worth of work is in place and has been invoiced. How much retention will have been deducted by the client based upon the following three scenarios?
    a. 10% retention rate
    b. 5% retention rate
    c. 10% retention rate until the project is 50% complete and no further retention held thereafter.

11. Not discussed in this chapter but why might a project owner or architect tie the monthly pay request process to the as-built drawing process? You may want to look ahead to Chapter 16 for this question.

15

Accounting for change orders

Introduction

Addenda modify the scope of the project before it is bid and change orders accomplish the same thing but after contract award. In addition to modifying the scope and terms and conditions of the agreement, a change order usually impacts cost and/or schedule. Change orders may be additive, if they add scope of work, or deductive, if they delete work items. Since they occur on most construction contracts, managing change orders is an important project management and cost engineering function. This chapter discusses change order management and pricing and the interplay of change orders between the jobsite and home office accounting operations.

Change order sources

Change orders arise from a variety of sources or causes. Four of the most common are elaborated on in this section. Most change orders originate due to *design errors*. These can arise for a variety of reasons, including:

- Architects and engineers may not have sufficient funds, resources, or time to do a complete design;
- Clients may contract with the engineers (such as mechanical or electrical engineers) directly and not through the architect. This may result in multi-discipline documents that are not fully coordinated;
- A complex project such as a hospital; and/or
- People make mistakes, including architects, engineers, and project owners.

Client-requested *scope* or *programming changes* are among the most expensive change orders but are relatively easy to process and gain approval for as this is something the project owner wants to add or change. Additional scope changes are usually accompanied with an architect-generated construction change authorization or construction change directive (CCA or CCD) with new or revised drawings and specifications.

Change orders which occur due to *uncovering unknown site conditions* can be among the most difficult type of change order for a client to accept as they were unforeseen or unanticipated. These generally result from inadequate site exploration prior to starting design. Hidden or latent conflicts or conditions are common in site work (for example, wet or contaminated soil) or remodeling (for example, mold or insect infestation) and may use up the client's contingency funds early in the project. When the contractor encounters hidden site conditions that adversely impact construction, the general contractor's (GC's) jobsite project team must 'promptly' notify the architect and client and provide the designer an opportunity to make an inspection; this is often through the use of a request for information (RFI).

Materials and equipment supplied by the project owner for installation by the contractor, or *not-in-contract (NIC) equipment*, result in a fourth category of change order if the products are not fully coordinated. Some construction consumers believe that by contracting direct with subcontractors and suppliers they save the general contractor's fee. This is common with items such as kitchen equipment, swimming pool subcontractors, exercise equipment, furnishings, custom casework, and landscaping on commercial projects. Problems may surface and the project can be disrupted due to the lack of coordination of client-furnished materials. These conflicts are unfortunately often not discovered until the materials first arrive on site. Conflicts may have been mitigated if the construction team had reviewed the client's purchase orders and vendor submittals and identified potential issues through the use of requests for information. Project owners may pay much more to resolve conflicts associated with NIC materials then they would have originally paid the general contractor to manage the work.

There are several other ways change orders may occur such as weather, changes initiated by the City when issuing the building permit, contractor and supplier requested material changes, and others. Most of the processes to develop and account and manage change orders will follow the same path as discussed in the balance of this chapter. The change order process begins with preparation and pricing of change order proposals (COPs). After the proposals have been approved, they are incorporated into the contract through formal change orders. Cost accounting and cost control systems must be continually modified to reflect approved change orders. If the contractor and the project owner cannot agree on change order pricing and time impacts, then the general contractor and its subcontractors may process claims which require a formal dispute resolution process to resolve.

Change order proposal process

The change order process can begin in a variety of fashions as indicated earlier. One common route is to start with a request for information due to a discrepancy discovered by the construction team, resulting in a cost proposal or request for proposal of subcontractors. The other route is through a project owner or architect-generated request for a change. In both cases the general contractor gathers all of the costs and backup, and forwards a change order proposal to the client or architect, and, if accepted, results in a contract change order (CO). The steps associated with processing COPs are discussed in this section. Claims are also briefly introduced later in this chapter. Claims are often the result of unresolved COPs.

Evergreen Construction Company
1449 Columbia Ave.
Seattle, WA 98202

CHANGE ORDER PROPOSAL

Project: Olympic Hotel and Resort, #422   PM: Chris Anderson
COP #   32   Date: 07/01/19

Description of work:

The GWB taping finish had been specified in specification section 091250 as a level three finish with a light 'orange peel' texture. In response to RFI 101, the project owner and architect desire a smooth level four finish with high-gloss enamel paint to facilitate cleaning.

Referenced documents:

RFI 101, Meeting notes 22.07, Phone Memorandum, GWB Subcontractor Change Request

COP Estimate Summary: Extended Cost:

1	Direct Labor:	hours @ $41/hour	(See att'd)	$0
2	Supervision:	hours @ $55/hour		$0
3	Labor Burden:	50% of labor:		$0
4	Safety:	2% of labor:		$0
5	Total Labor:			$0
6	Direct Materials and Equipment:		(See att'd)	$0
7	Small Tools:	1% of DL		$0
8	Consumables:	1% of DL		$0
9	Total Materials and Equipment:			$0
10	Subtotal Direct Work (Items 1 through 9):			$0
11	Subcontractors (See attached subcontractor quotes):			$50,000
12	Overhead on Direct Work Items:	(w/fee)		$0
13	Fee on Direct Work Items:	7% of DL & DM		$0
14	Fee on Subcontractors:	3% of subs		$1,500
15	Subtotal Overhead and Fee:			$1,500
16	Subtotal all costs:			$51,500
17	State Excise Tax:	0.75% of subtotal		$386
18	Liability Insurance:	0.95% of subtotal		$515
19	Total this COP # 32			**$52,015**

This added work has an impact on the overall project schedule, the extent of which cannot be thoroughly analyzed until after the change is incorporated into the contract and the work has been completed.
Please indicate acceptance by signing and returning one copy to our office within five dates of origination.

Approved by:

Sandi Chambers   07/06/19

Northwest Resorts, LLC   Date

Figure 15.1 Change order proposal

A change order is a formal contract-revising vehicle whereas a change order proposal is a softer and more negotiable document. An owner-initiated COP request provides a description of the proposed change in the scope of work and requests that the contractor provides an estimate of additional or deductive costs. If the proposed change extends the project duration, the contractor also requests additional time. A contractor-initiated COP generally results either from a differing site condition or from a response to a request for information. The contractor's proposal describes the change to the scope of work and provides any proposed adjustments to contract price and time. Figure 15.1 reflects a COP from Evergreen Construction Company for changing the level three, gypsum wall board (GWB) taping finish, to a level four finish throughout the hotel case study project. COPs do not modify the construction contract; only a formal contract change order accomplishes this. An additional COP example is included on the eResource for a change in the concrete foundation walls, including quantity take-off (QTO) and direct work pricing recap sheets.

Subcontractors account for 80–90% of the work on a typical commercial construction project. Subcontractors therefore initiate the majority of the change order proposals. Well-written requests for information and thorough and timely submittals are methods for surfacing discrepancies and therefore are part of the active quality control process. The project manager (PM) or project engineer/cost engineer should research a changed condition and forward the question or submittal to the architect. In this case the GWB subcontractor had processed its submittal indicating a level three taping finish, including a light texture, which was as specified, but the architect intended a level four finish and a smooth high-gloss paint which allows for easy cleaning. The architect changed the level of finish upon submittal return.

If the architect does not request a change order proposal from the contractor when responding to a request for information or submittal, the subcontractor must notify the general contractor regarding cost and schedule impacts from the architect's response. If the GC's team believes a change order is warranted, they will send a change order pricing request to the subcontractor. The project manager or jobsite cost engineer uses the subcontractor's response to the change order request to prepare the GC's COP similar to the one illustrated in Figure 15.1. COPs should include at least the following:

- The format of a COP cover page will be very similar to an estimate summary;
- Each COP is individually numbered;
- A detailed description of the work proposed is included;
- Direct labor hours, wage rates, and labor burden are itemized;
- Direct material and equipment costs should be backed up with detail;
- Subcontract costs are backed up with quotes from the individual subcontractors;
- Markups as established in the contract agreements are added; and
- The COP provides a line for the project owner to sign approval.

COPs should be prepared in a manner to facilitate expeditious approval from the project owner and architect. All relevant supporting documents should be attached. Subcontractors should provide the GC with the same level of detailed backup the owner asks of the GC. All COPs should be tracked with a log similar to an equipment ledger, value engineering, or an RFI log. Some of the types of COP backup documents which should accompany every change request include:

- RFIs and/or submittals which initiated the change;
- Relevant drawings, sketches, specifications, and/or photographs which depict the change;

- Estimating backup including quantity take-offs and pricing recap sheets;
- Subcontractor and supplier quotations; and
- Other supporting documents such as letters, phone memos, meeting notes, and/or daily diaries.

If COPs are properly prepared by the subcontractors and the GC, and the project owner accepts the COP, the formal change order process is relatively straight forward. Often, however, there will be questions related to the COP request and subsequent negotiations. This is to be expected and is part of working as a team. The cost proposal is usually negotiable. The project management team is not taking a hard line approach at this time. A weekly COP meeting outside of the regular owner-architect-contractor construction coordination meeting is a good way to discuss and resolve change issues. Following are two examples of subcontractors' approach to change orders; one successful and one not as successful.

Example One: Both of these subcontractors worked on the same biotechnology project. The electrical subcontractor had not done a lot of work in the high-technology arena, but they low-bid this project in hopes of negotiating future work. The project manager was a former superintendent and an electrician by trade. He filed requests for change orders for every minor issue, generating over 125 change order proposals, some of them for only $100. This frustrated the client who managed the project from out of town and visited the site only once monthly. The client fought with the subcontractor on almost every change and the subcontractor eventually filed a post-project claim. The electrical subcontractor did not break into the very tight-knit biotechnology construction market.

Example Two: The mechanical subcontractor also bid this project lump sum but had extensive biotechnology experience. Their project manager took a big picture approach. He did not charge for minor discrepancies. The client eventually generated a very major change in scope, adding and revising several pieces of mechanical equipment both on this project and in an adjacent facility. These changes amounted to over $600,000 worth of mechanical work. Many of the equipment changes were performed off-shift to minimize the impact to the client's employees and laboratory experiments. The client was very thankful. The mechanical subcontractor is still in the building doing time and materials (T&M) work years later.

Change order proposal pricing

Change order proposals cannot be processed for payment until they are approved and incorporated into the contract via formal contract change orders. The easiest way to obtain payment is to be realistic with respect to pricing on direct and subcontract cost estimates. Overly inflated prices will only delay the process. Quantity measurements generally are verifiable and should not be inflated. Wage rates paid to craft employees are verifiable and should not be inflated. Subcontractor quotes should be passed through as is without adjustment (unless incomplete) from the general contractor to the client. The subcontractors and suppliers should practice the same procedures with their second- and third-tier firms. Labor productivity rates should be derived from pre-approved resources. Material prices should be actual and verified with invoices or quotations. Any deviations in the previous suggestions may damage team-build relationships and trust amongst the parties and/or delay COP approvals. An effective approach to pricing changes is for both parties to pre-agree to utilize a published estimating database, such as from RS Means.

After all of the direct work and subcontracted work is totaled on the change order proposal, the general contractor will add a series of markups to the estimate. There are many percentage markups which may be added to the bottom of a COP including: labor burden, consumables, small tools, fee, potentially jobsite general conditions, contingency, liability insurance, bonds if applicable, and others. These percentage add-ons are often cumulative and their total effect on the direct cost of the work can be substantial. Markups, or percentage add-ons, are utilized by some GCs and subcontractors as fee enhancement and recovery for other indirect costs. Clients often get frustrated with these add-ons. They do not understand why they have to pay more for the change than simply the direct costs. Most of these markups are required, and sometimes it is just the presentation that makes them difficult to accept. An alternative approach is to figure the exact impact for each item, but this can be difficult for the contractor to account for and the project owner to understand and accept.

As discussed throughout this book, home office overhead and profit, or fee, is a very important cost add-on and is closely monitored by the home office chief executive and financial officers. Commercial general contractors will receive a 3–7% markup for fee. Civil, residential, and smaller commercial projects will see up to 10% fees. This rate is usually the same percentage fee that was used on the original estimate and is stated or allowed in the contract. Evergreen Construction Company (ECC) and Northwest Resorts (NWR) agreed to a 7% fee markup for additive direct work and 3% for subcontracted work and inserted that into their A102 contract, Article 5.1.2. There was no deduction of fee for reductions in scope, which is often the case, unless there is a major scope change, such as the elimination of a parking garage floor. Home office overhead is typically included in a general contractor's fee and is not a separate add-on for GCs.

Most construction contracts also do not allow the GC to charge for additional jobsite general conditions on change orders. The exception is if schedule extension can be proven, then some jobsite general conditions line items may be allowed. In order for the jobsite project team to prove that additional general conditions and/or time are warranted, they need to provide documentation to back up the time and costs, such as a very detailed construction schedule. Contractor attempts to recover for additional jobsite general conditions, time extensions, home office general conditions, and/or impact costs are often seen as contentious from the owner's perspective and are typically rejected in the change order proposal process, in which case they could result in a contract claim. Development and management of jobsite general conditions estimates was the topic of Chapter 5.

Specialty contractors often charge a higher fee on change orders than will a general contractor because their volume of work is less and their direct labor percentage is higher. Direct labor is an estimating risk for subcontractors as well, even on change orders. Subcontractors may receive a 10% fee and an additional 10% for overhead. Fee and overhead rates allowed on change orders will depend upon how many other markup items are included in, or are in addition to, the fee and overhead markup. Contract language should define all of this.

Negotiations and approvals

A formal contract change order is utilized to modify the construction contract after the contract is executed. The contract change order is generally prepared by the architect. If the architect and client and the general contractor have negotiated a mutually acceptable adjustment in the contract price, schedule, or both through the use of the change order proposal process, a contract change order is executed modifying the terms of the contract. The supplemental or special specifications of most prime contracts contain the change order procedures. ConsensusDocs form 202 and AIA document G701 are common formal change order coversheets and preferred by many contractors, clients, and architects. Because Resorts International, the holding company for Northwest Resorts, which is the owner for the Olympic Hotel and Resort case study project, builds and owns hotels across the country, they have their own customized formal contract change order form. An example contract change order incorporating the gypsum wall board COP and others is included as Figure 15.2.

CONTRACT CHANGE ORDER

Where the basis of payment is a guaranteed maximum price

Client: Northwest Resorts, LLC
Address: 1001 First Avenue, Seattle, WA 92012, (206) 392-5336

Contractor: Evergreen Construction Company
Address: 1449 Columbia Avenue, Seattle, WA 98202, (206) 447-4222

Project:	Olympic Hotel and Resort	Project Number:	422
Change Order Number: 2		Change Order Date:	8/21/19

Original Contract Price:		$24,530,015
Previously Approved Change Orders:	Through Change #: 1	$ 75,050
Previous Contract Price:		$24,605,065

Incorporate the following agreed change order proposals. See attached COP log.

13	Unsuitable soil removal	$ 27,550	
24	Building Permit Issue	$ 0	
25	Add ATB Asphalt for Winter	$ 44,567	
28	Extend Roof Warranty to 25 Yr.	$ 20,102	
32	Level Four GWB Finish	$ 52,015	
Total added/reduced/unchanged this CO:		$117,245	$ 117,245
Current and revised Contract Price:			$24,722,310

Previous Contracted Schedule Completion:		7/10/2020
Net change per this CO: 0	Days added/reduced/unchanged:	0
Current and Revised Contracted Schedule Completion:		7/10/2020

Contractor's Approval:	*Chris Anderson*	Date: 8/21/19
Architect's Approval:	*Stan Zimmer*	Date: 8/22/19
Client's Approval:	*Sandi Chambers*	Date: 8/22/19

Figure 15.2 Contract change order

A formal contract change order can be used either for one large change order proposal or a group of COPs. Some clients choose to issue monthly formal change orders incorporating all COPs approved during the month. This is a good process for the general contractor if the change order is executed before the monthly pay request is drafted, allowing the CO to be invoiced this month. COPs do not modify the terms of the contract, and are not added to the schedule of values for payment purposes until incorporated into the contract by formal contract change order. Because the contract often does not allow payment of COPs until they are formally incorporated into the contract and added to the pay request SOV, expediting the formal contract change order is of high importance to the construction team. The contract documents should be annotated with all scope changes contained in each change order. The project manager should walk the formal change order through, gaining approval signatures from the architect and the project client. Once the change order has been signed by the architect, client, and contractor, the contract scope, price, and time have been modified. There often is language in both the COP and formal contract change order that indicates once the GC has signed they agree with the adjustment in price and time and will not submit an additional claim for this scope. The GC will incorporate similar language in their changes to subcontractors and suppliers.

Incorporation into cost control and accounting control systems

After the general contractor's contract has been modified, the project manager or cost engineer will modify appropriate subcontracts and purchase orders for major material suppliers. Only then can the subcontractors and suppliers begin their pay request process for the changes.

After the formal contract change order has been executed, the project manager will generate journal entries to the cost accounting system, increasing (or decreasing) the total contract value, adjusting individual estimate line items for approved changes of scope, and modifying percentage markups such as insurance and fees. These changes to the home office accounting system will be done through the use of journal entries, often initiated by the jobsite cost accountant for approval by the PM. These journal entries do not need to 'balance' as did changes made to the cost control system during the early estimate correction stage of the cost control cycle. The net effect should equal the total value of the approved change orders.

The cost engineer will assign new cost codes to added scope and assist the foremen with work package development. If the changes are being performed on a time and material basis, it is important that only true costs be charged to those cost codes. The client trusts the contractor to be fair when performing work T&M and the construction team can damage relations if mis-coding occurs. Some large open-book projects may have the on-site owner's representative actually initial on time sheets and material invoices associated with change orders. In some instances, if the parties cannot agree on a price before the work commences, the contract may allow the client to direct the contractor to proceed either on a T&M basis or on a price 'to-be-negotiated' basis. In either case it is important that the costs be tracked accurately. Actual tracking of some costs are also accomplished inadvertently through the use of the superintendent's diary when he or she notes manpower on site and work accomplished, materials delivered, and equipment utilized. The equipment ledger will also track if new equipment was used for changed conditions.

There are two methods to incorporate change orders into the pay request process. The schedule of values had been established and approved at the beginning of the project. Each line item affected by changes may be modified on a monthly basis when changes are approved. The problem with this method is it is difficult to keep track of the original estimate and melding invoices for original scopes with changed scopes. The other system is to add line items at the bottom of the SOV listing approved contract changes and invoicing against them directly. The problem with this method is it is difficult to track the revised subcontract and purchase order values and match monthly invoices with lien releases. Evergreen Construction Company utilized the second approach in its SOV and monthly invoices with Northwest Resorts on the Olympic Hotel case study.

Unresolved change orders

Some regard all change orders as 'claims' but claims are not necessarily the same thing as a change order proposal or a formal contract change order. Claims often stem from a COP that has been rejected by the client, or added scope that was forced on the contractor. Many claims arise late in the construction process, after contractors have realized that they did not achieve their anticipated estimate or schedule. For this reason, the contract will stipulate time frames for when a claim must be filed, for example within 21 days of occurrence. The value of a claim is also customarily much larger than a COP. A COP may be for $3,000 on the Olympic Hotel case study project for a change to hotel room signage, whereas the contractor may file a claim for $350,000 or more due to a two-month schedule extension caused by the owner or an unusually cold winter. Validation of actual costs incurred plays an important role in claim settlements. It is difficult to argue against 'actual' costs on changed conditions versus 'estimated' costs and it is through the use of accurate cost coding and recording by the cost engineer that this is accomplished.

The method to resolve claims is described in the prime contract agreement. The least expensive and most expedient solution is prevention which requires all parties to be diligent about their communications and resolve issues at the lowest level possible, which is at the jobsite, before they have escalated back to the home office. If a conflict arises, the parties are encouraged to negotiate its solution, also at the jobsite level. The change order proposal process described earlier is the preferred less expensive path. If a change is not resolved between the parties at the project level, the claim resolution process will likely follow these discourses, in order of least to most expensive and time to resolve: mediation, dispute resolution boards (DRB), arbitration, and litigation. Claims and dispute resolution are not directly a cost accounting topic, but good accounting practices can facilitate either claim prevention or successful claim resolution.

Summary

Change orders are part of the construction process and occur on almost every project. The jobsite management team, including the owner, architect, and general contractor, needs to deal with changes in the most expedient and fair manner possible. Change orders arise from a variety of sources including discrepant design documents, scope additions, differing site conditions, and not-in-contract equipment. Change order proposals may be initiated either by the client or by the GC. An owner-initiated proposal requests the project team provide an estimate of the cost and schedule impact. A GC- or subcontractor-initiated proposal describes the proposed change in the scope of work and requests an acceptable adjustment to the contract price and/or time. The project team maintains a COP log to track all COPs whether generated by the client, GC, or subcontractors. Once the GC and the client have negotiated a mutually agreeable adjustment in contract price and time, a formal change order is executed modifying the contract. A claim differs from a COP or contract change order in that it is a request for time and/or cost by the contractor, and the client has not yet agreed to accept, and will be resolved according to the dispute resolution process defined in the prime contract agreement. The most efficient way to process and attain approval for change orders is clear and timely communications between the parties, completely open-book accounting, fair pricing, and good bookkeeping at the jobsite level.

Review questions

1. What is the difference between a COP and a formal contract change order and a claim?
2. Which method of dispute resolution will the parties follow?
3. How much backup should a PM or estimator or cost engineer include with a COP?
4. What is the preferred timing of incorporating formal contract change orders and why then?
5. What would happen to the GC's PM if the client discovered pricing in the COP had been falsified or inflated?
6. When would it be beneficial for a contractor to perform change order work on a fixed price basis?
7. When would it be beneficial for a contractor to perform change order work on a T&M basis?

Exercises

1. Other than the four typical change clauses described in this chapter, how might a changed condition warranting additional cost and/or time occur?
2. Prepare a flow chart identifying a discrepant condition, such as an incorrect soap dispenser in ten hotel room showers identified by NWR's owner's representative during a walkthrough, the direction to repair the problem from the architect, ordering and delivering materials, performing the work, preparing a COP and CO, and finally paying the GC and the material supplier.
3. If the GC cannot get a subcontractor to price a change fairly, what should they do?
4. Prepare a COP for the Olympic Hotel case study project for the addition of restaurant kitchen equipment. Include the following costs:
   a. Miscellaneous architectural subcontractors totaling $7,500,
   b. 100 hours for ECC's carpenter labor,
   c. Kitchen equipment purchase and installation at $250,000 by a subcontractor,
   d. Plumbing changes amounting to $15,000,
   e. Electrical work priced at $35,000,
   f. Mechanical exhaust for the grease hood costing $25,000,
   g. One week total schedule extension.
5. Have you experienced a successful change management operation and/or an unsuccessful one? Why do you feel this was the case?

16

Financial close-out of the construction project

Introduction

As construction projects near physical completion, contractors transition into a close-out phase. The jobsite team should develop a project close-out plan to manage the numerous activities involved in closing out the job. Just as start-up activities are essential when initiating work on a project, good close-out procedures are essential to timely completion of contractual requirements and receipt of final payment. Not only is efficient project close-out good for the general contractor (GC), but it is also good for the client. All contractors want to close out a job quickly and move on to another project. Minimizing the duration of close-out activities generally enhances profit, as it limits jobsite general conditions costs and facilitates timely receipt of final payment and release of retention. Efficient close-out and turnover procedures also minimize the contractor's interference with the client's move-in and start-up activities. Contractors may lose client relationships and damage their reputation in the industry because of slow close-out processing.

Close-out involves completing all physical construction work and assembling all paperwork documentation required to contractually close out the project. The prime client-GC contract will address requirements for substantial completion and final payment. The AIA A102 contract utilized on the hotel case study outlines the close-out activities required to process final payments in Article 12.2. Superintendents generally are responsible for accomplishing construction field work and physically closing out the project. The project manager (PM), with the assistance of the jobsite cost engineer, is responsible for the paper close-out, including all of the associated financial and accounting aspects. This chapter discusses elements of close-out implementation including: construction close-out, project management close-out, financial close-out, and as-built estimates.

Construction close-out

The work required to complete construction and close out physical aspects of the project are the responsibility of the project superintendent. There are many physical construction close-out activities including:

- Complete construction in accordance with contract requirements;
- Punch list walkthroughs, documentation, and resolution of outstanding items;
- Start-up, commissioning, and training;
- Certificates of substantial completion and occupancy;
- Client move-in; and
- Physically moving off the jobsite, or demobilization.

The jobsite team, headed by the superintendent, continuously inspects completed work and identifies deficiencies during construction of the project as part of an active quality control program. In-process *punch lists* are one of the tools a general contractor will utilize to reduce the size of the final punch list which ultimately saves all parties money. When the project is nearing completion, the GC requests a walkthrough inspection from the owner and architect. The list of deficiencies identified during this inspection is known as the punch list, or formal punch list. The punch list usually is developed by the architect, but some clients prepare their own. Some consultant design team members, such as mechanical or electrical engineers, may also develop punch lists. The best method is for all parties who are interested in inspecting the project to gather and walk the jobsite at the same time and one formal punch list is developed.

Timing of the formal punch list walkthrough is critical. It should be early enough to allow time for corrections, but not so early as to mix incomplete base contract work with repairs of completed work. If there still are basic construction activities to complete, additional damage may be incurred that was not listed on the original punch list and it therefore would require revisions. Conversely, if the project team has waited to develop the punch list until after the client has moved in, it is difficult to determine who did what damage.

The general contractor should take no more than one month to repair all of the work identified on the punch list. All items on the punch list must be corrected before the project can be considered 100% complete. If the process takes too long, the subcontractors will likely have demobilized and it may be difficult to get them remobilized back on the jobsite. The punch list should be signed off by each subcontractor and the GC as deficiencies are corrected. A copy of the completed punch list should be sent to the architect. The architect may then wish to re-visit each punch list item and verify its completion or may perform spot-check audits. The certificate of substantial completion is then originated by the architect.

All of the mechanical, electrical, and plumbing (MEP) systems and equipment need to be *started and tested* to assure they are performing properly before turning over the completed project to the client and their facility maintenance personnel. The MEP subcontractors usually take the point on testing and must follow the requirements in the specifications as well as those dictated by equipment manufacturers. These tests are scheduled with the GC's superintendent and the client and often the design engineers will attend. Once complete, a test certificate will be generated and all present will sign. These reports are then customarily included in the operation and maintenance (O&M) manuals as discussed later. The heating, ventilation, and air conditioning (HVAC) systems will also require to be balanced – sometimes by an independent third party. Many projects today also are commissioned either by the contractor, the client, or again by an independent third party. All of these HVAC balance and commissioning reports will also be included in the O&M manuals.

The *certificate of substantial completion* is issued by the architect, or engineer in the case of a heavy-civil project. This important certificate indicates the project is sufficiently complete such that it can be used for its intended purpose, but there still may be some minor deficiencies that need to be completed, hence 'substantial.' All items on the punch list may not have been corrected but the architect agrees that the project is ready to be used. A copy of the corrected or updated punch list should be attached to this certificate. Contractual completion is the date by which substantial completion is achieved. Issuance of the substantial completion certificate ends the GC's liability for liquidated damages (LDs) and signifies the start of the warranty period.

The authority having jurisdiction (AHJ) over the project site, which is likely the City, County, State or the Federal Government, issues the *certificate of occupancy* (C of O). This is the same agency that issued the original building permit. The C of O signifies all code-related issues have been accounted for and that the building is approved and safe to use. It is often a formality which follows completion of all of the various inspections and subsequent approval and sign-off of all other construction permits. The most critical aspects of the certificate of occupancy are usually inspection and approval from the fire marshal for life-safety issues along with the elevator inspection. The project may have minor deficiencies that are not related to life-safety and still be approved for occupancy. For example, one window blind in hotel room # 405 may not yet be installed or corrected, but it is safe from a life-safety perspective and the balance of the hotel is still usable and built to code. Some public agencies will issue a temporary certificate of occupancy (TCO) which allows the client conditional use of the building for a stated period of time, say six months, while other non-critical work such as recreational room exercise equipment or some interior finishes may be completed at a later time. In order to obtain contractual close-out, the GC's jobsite team needs to obtain both certificates of completion.

Most project owners are anxious to move into their new projects as soon as the authority having jurisdiction allows them to be occupied. *Joint occupancy* occurs when the client accepts and occupies a portion of the building while the contractor is still working in another portion. This may be dictated by the client's need to begin business in the new facility, host a conference, or the client may want to begin moving equipment and furniture into the building. Joint, dual, or conditional occupancies often are unavoidable and may be undesirable if not managed properly. The project team may have problems differentiating the client's cleanup and routine maintenance from construction punch list and warranty work. It is better for all parties if occupancy is delayed until both certificates of completion have been obtained and all items on the punch list have been corrected. If the GC and the client are to have joint occupancy, both parties have to work together to establish procedures to accomplish each other's goals; all of which needs to be identified in the contract.

Commercial estimators seldom put a line item in their original estimate for either mobilization or *demobilization*. Demobilizing, or physically moving off the site, involves considerable work and can be expensive, especially to the general contractor. At that time, there may not be funds that can be dedicated to a foreman and a small crew to clean up the site. The question always arises between the GC and its subcontractors as to whose garbage and surplus materials remain. Similar to the discussion regarding joint occupancy, garbage accumulation may become mixed between the construction crews and the client, as the client is also in the process of moving in.

Demobilization involves closing the project office and physically removing all contractor-owned equipment, files, and personnel from the project site.

Project management close-out

Close-out starts for the project management team during the project start-up phase by reviewing the contract and specifications for close-out requirements. This ideally happens early so that close-out requirements are delineated in each subcontract agreement and major purchase order. The project manager (PM) or project engineer (PE) should make a list of what they think needs to be done and then ask the design team for verification. Some of the major items the project team will submit during the close-out process include as-built drawings, operation and maintenance manuals, test reports, extra finish materials, and signed-off permits.

During the course of construction, actual dimensions and conditions of the installed work are noted on the contract drawings. These drawings then reflect the as-built conditions which therefore should be more accurate than were the original design drawings. The project manager and jobsite project engineer should collect all of the *as-built drawings* from the subcontractors. The general contractor develops as-built drawings for civil, architectural, and structural work; MEP subcontractors prepare and submit their own as-built drawings to the GC's PE. Mechanical, electrical, plumbing, and civil installations are very important because they include hidden systems that would cause severe damage if cut or may need to be accessed during a future remodel.

The foreman or assistant superintendent who oversaw construction is the most appropriate project team member to mark up the as-built drawings. The as-built drawings should be submitted to the client or designer using the same procedures as shop drawings were submitted. Many projects which utilized building information modeling or computer-aided design by the architectural and engineering teams may also require contractors to record as-built conditions electronically. An additional specialized project engineer may be added to the general contractor's jobsite staff and general conditions estimate if electronic as-built drawings are required.

Operation and maintenance manuals are used to gather manufacturers' data regarding operational, preventive maintenance, cleaning, and repair procedures for all equipment as well as many of the architectural finish materials. This information should be collected from all the subcontractors and suppliers and organized in several sets of three-ring binders for the client's permanent service record. The contract specifications will describe the format and organization of O&M manuals. It is a good procedure to process the O&M manuals also as a submittal and request architect approval. This is sometimes required and always a good idea. Some designers will request draft copies be submitted for comment prior to submission of the final copy. Similar to as-built drawings discussed earlier, the O&M manuals may be required to be submitted electronically.

Throughout the course of construction, materials and systems are started up and tested and *test reports* document the results. This includes *balancing and commissioning reports* for the mechanical systems. All of these test reports may be bound as a section in the O&M manuals. The project engineer will help with collection and filing of these reports and final assembly of the O&M manuals.

Specifications require *extra material quantities*, say 1–3% of various architectural materials, be supplied to the client upon completion. This would include materials such as paint (of each color used), ceramic tile, carpet, and ceiling tile. This assists the client with future repairs and remodels.

Construction craftsmen sometimes incorrectly use up these materials for punch list or change order work and contractors run a danger of being short at the time of turnover. Extra finish materials should be kept in secure storage or turned over to the client once received.

Many subcontractors were required to obtain their own building *permits* from the authority having jurisdiction. This often includes firms such as electrical, elevator, and fire protection contractors. Once complete, their permits must also be signed off by the AHJ. This indicates that the subcontractor's work was performed in conformance with code requirements. Signed-off permits may also be included as a section in the operation and maintenance manuals, similar to other close-out documents. The jobsite management team also may be required to submit to the designer or include in the O&M manuals all interim *inspection reports* or signature cards received from the city or county throughout the course of construction. All of these items should be included in a close-out log.

Financial close-out

The financial close-out of a project is not the chief financial officer or the officer-in-charge's (CFO or OIC) responsibility. The general contractor's project manager is responsible to financially close-out the project with both the client and his or her subcontractors, but may obtain the assistance of upper-management as well as the jobsite cost engineer. The GC's ultimate goal is to receive its retention, and all issues related to financially closing out the project are necessary before the client will write the final retention check. Some of the items involved in financial close-out of a project include final change orders, final payment request, client audits, final lien releases, final fee forecast, and others.

Final change orders

Financial close-out for the general contractor's project manager includes negotiating final change orders with the client. The client wants to make sure that change orders will not continue to trickle in for months after they have agreed the project is complete and have accepted it for use. Change orders which may have been rejected or tabled earlier must now be resolved. Unresolved change orders tend to turn into claims as was discussed in the last chapter. The client should request a final change order from the GC, even at $0.00, with the word FINAL printed clearly at the top.

Before the general contractor's project manager can guarantee to his or her client that there will not be any further change orders, they must 'shake the tree' and be sure that all of their subcontractors and suppliers also do not have any pending cost issues. On any typical commercial construction project there may be 100 subcontractors and suppliers, so assuring that they all are satisfied and will not be submitting a latent change request or claim is a difficult financial close-out task. As in all areas of risk-management for the GC, they should ask the same from their subcontractors as the client asks of them. In this case that means the GC's PM and cost engineer should issue FINAL subcontract and purchase order modifications to their subcontractors and suppliers, which incorporate all of the remaining change issues and back charges, even again if written for $0.00. An example final subcontract change order for Evergreen's

---

Evergreen Construction Company
1449 Columbia Avenue, Seattle, WA 98202
206.447.4222

FINAL SUBCONTRACT CHANGE ORDER

Project: *Olympic Hotel and Resort*
Subcontractor: *Hoquiam Drywall*   Subcontract Modification Number: **003**

We agree to make the following change(s) in the subcontract scope of work:

Mod 1: *Incorporate lobby finishes allowance ($274,000 in base bid, mod @+$0)*
Mod 2: *Change from level 3 and light texture to level 4 finish (+$50,000)*
           Spec Section 092500, RFI 101, Sub proposal 1 and COP 32
Mod 3: *Final subcontract modification for no change in time or cost*

Subject to the following adjustment(s) to subcontract value:
    Original Subcontract Amount: $ 999,000
    Previous Subcontract Modifications (through mod *002*): $ 50,000
    Previous Subcontract Amount: $1,049,000

    Cost of this Modification Number *003*: $ 0
    Revised Subcontract Amount: $1,049,000

    Revised Completion Date: *No change*

Issued by/Contractor: Evergreen Construction Company
By: *Chris Anderson*, Project Manager   Date: *9/28/2020*
                  Authorized Signature/Title

Accepted by/Subcontractor: *Hoquiam Drywall*
By: *Jimmy Dunham*, President   Date: *9/30/2020*
                  Authorized Signature/Title

Figure 16.1 Final subcontract change order

gypsum wall board subcontractor on the Olympic Hotel and Resort case study is included here as Figure 16.1.

Final payment requests

The general contractor will prepare a final payment request for the client. Some GC project managers will include their last month's progress payment request combined with their request for retention. This might result in complications if the client is okay with the quality of the work completed in the last month, and is willing to pay that amount, but is not yet ready to release the GC's retention. If the two issues are combined with one request, and have one lien release with the totals combined, the client may hold back on the entire amount. This will result in a financial strain on the GC and its subcontractors. It is recommended that two separate requests for payment be prepared, one for the last month of work and one for release of retention, even if the two are submitted to the client at one time. This scenario is shown in the following examples:

Example One: Combined month 16 pay request with complete release of retention:

Month 16 work accomplished: $475,000 ($500,000 – 5% retention of $25,000)
+ whole project retention release of $1,500,000 (including their last $25,000)
Invoice #16: Total request = $1,975,000

Example Two: Month 16 pay request separate from retention request:

Invoice #16: Month 16 work: $475,000 ($500,000 – 5% retention of $25,000)
Invoice #17: Whole project retention release of $1,500,000

The advantage for the general contractor and its subcontractors with Example Two is they can now receive $475,000 to use to pay for direct materials and labor, hopefully within ten days of submission of invoice #16, whereas invoice #17 will be predicated upon completion of all of the other project management and financial close-out items discussed in this chapter.

Preparation of the final request for payment for the general contractor's project manager and cost engineer is predicated upon their receipt of final pay requests from subcontractors. Similar to the discussion of final change orders earlier, this can also be a daunting task. The client should ask for the word FINAL clearly delineated on the pay application, and the GC should do the same with its subcontractors and suppliers. Once paid by the client, the GC's project team should expeditiously release payments to subcontractors. The pay request process will be accompanied by final lien releases from all of the GC's vendors and preparation of the same for the client as discussed later.

Audits

Taxes and audits are the subject of upcoming Chapter 18. Lump sum projects are fairly simple in that clients cannot audit the contractor's books because they are closed-book, unless of course there was contract language which allowed this, but that would be the exception and not the rule. Lump sum projects can have change orders that are authorized on a time and materials (T&M) basis and those change orders may be subject to the client or its agent reviewing the contractor's cost accounting, but for the change order only.

Open-book negotiated projects are often subject to a final audit by the client. Northwest Resorts, LLC (NWR) employed Evergreen Construction Company (ECC) to build the Olympic Hotel and Resort case study project on a negotiated basis with a guaranteed maximum price (GMP) of $24.5 million. At the end of the project, when ECC is preparing to request release of their retention, NWR will customarily audit the contractor's books to make sure that the contractor spent what they claim to have spent and only charged costs to the project that were agreed to be reimbursable as defined in the contract. NWR must receive an acceptable report from its auditor (whether that be internal or outside certified public accountant), or they will not release ECC's retention, or at least not all of it. If the general contractor has employed open-book subcontractors, such as design-build mechanical, electrical, and plumbing, the GC may also audit the subcontractors' books and this will again be prior to the client's audit.

If the contractor is signatory to labor unions, the unions may also audit the contractor's books, but usually on an annual basis and not on a project basis, to verify that the contractor has paid its share of labor benefits. A sophisticated client again may request that the unions provide affidavits that their benefits have been paid to date, especially as related to their project, as the unions have a right to lien the project if unpaid. These union affidavits may be another section in the operation and maintenance manuals.

Final lien releases

The final change orders and final pay request processes are dependent upon several other prerequisites, but all happening at the same time. This is somewhat a 'chicken and egg' scenario. The processing of the final lien release is very similar. One item cannot happen until another is resolved, and that item is also pending another issue. Monthly conditional lien releases were discussed in Chapter 14, 'Payment requests.' At the end of the project, the client is looking for a FINAL or unconditional lien release from the general contractor. The conditional releases basically indicate that:

> *If the client pays the GC for a specified amount of money for a specified period of time, the GC will release its rights to lien the project for exactly that amount.*

The client is then protected from liens once that amount is paid. An unconditional lien release indicates that:

> *The GC has been paid and now completely and unconditionally releases its right to lien for the amount received.*

The owner will not release the general contractor's retention until all of the close-out requirements discussed in this chapter have been completed, including this final lien release. The GC is protected during the course of the project as they know they always have one or more pay requests coming, so that they have not completely given up all of their lien rights. But this is not the case with the final lien release. The client is basically asking for an unconditional release, even though the retention has not been paid. This author/project manager has simultaneously exchanged many unconditional lien releases for the retention check with the client. Lien releases can have complicated legal language and are usually drafted by attorneys. Contractors should check with their legal counsel before agreeing to sign a client's lien release form.

The general contractor's project manager again places the same responsibility on the subcontractors as the client has placed on him or her. The GC should not provide the client with an unconditional lien release until it has received all of the final lien releases from all of its subcontractors, and their subcontractors. The GC should reach as far back as they can to ensure that third-tier subcontractors and their suppliers have all been paid. Experienced owner's representatives will also ask the GC to provide the subcontractor lien releases, along with its own, and include them in the operation and maintenance manuals.

Retention release

Release of retention is the ultimate goal for the general contractor as well as their subcontractors. As stated in Chapter 14, retention use to be equal to 10% of the cost of the work and in many cases is now 5%, which approximately equals a contractor's fee. Until the retention has been paid, the contractor has not received its fee. But GCs who subcontract out the majority of the work will also have held a corresponding amount of retention from their subcontractors as the client did of them. Therefore the GC has been receiving portions of its fee, and only the retention held for their direct labor and material and equipment and indirect costs is outstanding. At-risk construction managers, which subcontract out the entire project, have only 5% held on their indirect labor. It is the subcontractors who are helping the client to finance the project due to being paid their monthly pay requests on a pay-when-paid basis, as well as retention held. Third-tier subcontractors are impacted even more.

Before the retention is released, all of the topics discussed in this chapter need to be 100% complete. Any one item can hold up release of retention. This starts with construction close-out procedures such as the punch list and certificates of completion. All of the project management paper close-out activities must be assembled and submitted to the architect and the client including as-built drawings, operation and maintenance manuals, and several others. The jobsite cost accountant will assist the project manager with processing several very important financial close-out activities such as final change orders, final pay requests, and final lien releases.

The actual release of retention should be by separate check as noted earlier. This is a milestone event for the general contractor. Once received, the GC should make sure that all of the close-out activities of its subcontractors have been resolved and pay them expeditiously as well. The GC relies on the financial well-being of its subcontractors to not only get them through this project, but to be solvent and provide them with a competitive bid on the next project.

Warranty management and accounting

One additional phase that occurs after all of the construction phases have been completed, including the close-out phase, is the warranty period. The specifications and the contract will indicate how long the project is warranted, which is usually one year. Several building systems and equipment are warranted longer, such as the roof for 20 years and the elevator for five years. During that first year of warranty, the general contractor will spend funds responding to warranty call-backs, which can range from leaking faucets to drywall cracks to dead landscaping. It is important for long-term relations with the client that the GC and its subcontractors respond quickly and repair any deficiencies. As stated at the beginning of this chapter, a poorly closed out project can have an effect on a contractor's reputation and this is similar with warranty management.

For lump sum projects, the contractor may have allowed a small amount of money in its jobsite general conditions estimate for the warranty period. In most cases, the general contractor is relying on subcontractors to take the majority of the risk for warranty call-backs, but still the GC must manage the process. If the GC does not spend what they allowed, then they potentially improve their fee. If they over-spend their warranty allowance, then they will eat into their fee, just as an extended close-out phase will eat into the fee.

On negotiated open-book projects, the contractor will also have a general conditions line item for warranty. Evergreen Construction Company included $10,400 for warranty repairs as shown in the expanded general conditions estimate on the eResource. During the final audit the amount of money in that line item will come up for discussion. Can the GC really completely close out the project if there is still money set aside for warranty? That money may be just the right amount, in which case there is not an issue. But if there is money left over after the warranty period, it should belong to the client; again contract and savings split issues will determine this. If the contractor requires more funds than estimated, then other savings which may have already been returned to the client during the close-out financial audit would be necessary to cover the short-fall. Open-book negotiated projects with a guaranteed maximum price have protected the client on the high side for cost over-runs, but line-item over- and under-runs within the contract are to be expected and are available for the contractor to manage. So if the contractor under-ran its door and hardware estimate line item by $9,200, but over-ran its waterproofing estimate by $8,800, the two would nearly balance out. The same could be said for warranty estimates. The contractor does not want to take a risk on a warranty cost over-run yet the owner does not want the contractor to realize a windfall profit if they under-ran that estimate as well. Usually a compromise is established and the contractor agrees to track those costs separately, as an allowance, and if there is a significant over- or under-cost occurrence, which can be substantiated with time sheets for labor and material invoices, then the two will agree to resolve it a year from close-out.

Final fee forecast

Once the final monthly pay request and retention checks have been received, and all of the subcontractors have been paid, the project manager and cost engineer must perform a final cost and fee forecast. With each of the monthly forecasts the PM was predicting what the final fee would be at the end of the project. As the project progresses past buyout and through the direct labor portion of the project, the PM should become more accurate and confident in his or her forecasts. The OIC and CFO rely on accurate forecasts from each project to add up to a total income projection for the firm for the fiscal year. If Evergreen's PM has been forecasting a fee of $1.2 million for 15 months, but now at the end of the job surprises the front office with a final fee of $900,000, they will not be pleased. Conversely if the PM was forecasting overly pessimistic and the project came in with a final $1.6 million fee, although the front office would be happier than the previous scenario, they still would not be pleased. Consistent income projections are used to negotiate bonding and banking resources and to report to equity partners. Had the home office known the project was going to come in with a larger fee, they may have made other business decisions such as increasing bonding capacity and securing more work, or leveraging their equity to purchase additional equipment or expand into another market. The home office does not expect the PM's monthly forecasts to be exact from month to month, and in fact if they were exact one would wonder if the PM was backing into the fee forecast. Variations are acceptable, as long as they are manageable and the jobsite team can explain them.

One additional aspect of close-out that is often associated with the final fee forecast is an internal lessons-learned report generated by the project team. This is a useful tool that can help better prepare individuals and the company for future projects, possibly with this client or architect, or possibly on similar types of work. The lessons-learned report is basically a self-diagnosis of 'what worked and what didn't and why, and what we can do better on the next job.' A powerful addition to that report is an assessment of the GC's subcontractors and suppliers. Other PMs and staff specialists such as estimators and procurement specialists can benefit from a candid report card on subcontractor performance.

As-built estimate development and database input

The as-built estimate should be maintained throughout construction or prepared near the completion of the project but not months after close-out. Considerable work went into tracking actual costs. This is valuable input to the construction firm's ongoing ability to improve its estimating accuracy. Input of the as-built estimate into the estimating database is necessary, if the database is to be kept current. Many project managers or jobsite cost engineers will simply input actual costs alongside of the original quantities to the company's database. This is better than no input at all, but the most accurate historical cost data is created by combining actual direct labor hours and actual material costs with actual installed quantities, as exhibited in the following examples:

Example Three:

Estimated: 120 kitchenettes @ 20 man-hours per each (MH/EA) = 2,400 MHs
2,400 MHs @ $39/hour (HR) = $93,600 labor cost
Spent: $88,000 to install 122 kitchenettes (missed two reception areas), therefore:
$88,000/$39/HR = 2,256 hours/122 kitchenettes = 18.5 MH/EA labor productivity

---

Example Four:

Estimated: 200 window blinds @ $200/EA = $40,000 material cost
Spent: $42,000 to purchase 190 blinds (no blinds in pool room), therefore:
$42,000/190 EA = $221/blind unit price to purchase

If actual material costs and hours had not been factored in with the actual installed quantities, the historical as-built database would have been skewed. The contractor's estimating database is a compilation of all of their past projects. If averages only are looked at in the database, then extremely expensive or economical projects or line items or cost codes may distort the average. For example, looking at the window blind example in Example Four, if the contractor had several projects which ranged in window blind prices from $150 to $250 per each, then they would feel comfortable with the $200 average. But if two projects were factored in which had wood slats and

Table 16.1 Estimating database

<table>
<tr><td colspan="4"></td><td colspan="5">Evergreen Construction Company</td></tr>
<tr><td colspan="4"></td><td colspan="5">**Estimating Database**</td></tr>
<tr><td colspan="4"></td><td colspan="5">(Abbreviated)</td></tr>
<tr><td>Job Number:</td><td></td><td></td><td>109</td><td>114</td><td>332</td><td>410</td><td></td></tr>
<tr><td>Project Type:</td><td></td><td></td><td>Retail</td><td>Mid-rise</td><td>Middle</td><td>Spec</td><td></td></tr>
<tr><td>Category</td><td>Cost Code</td><td>Units</td><td>T.I.</td><td>Apartment</td><td>School</td><td>Office</td><td>Average</td></tr>
<tr><td>Buy 1 x 4 Fir Trim</td><td>062214</td><td>LF</td><td>$2.20</td><td>$1.80</td><td>$1.75</td><td>$4.00</td><td>**$2.52**</td></tr>
<tr><td>Install Kitchenettes</td><td>641100</td><td>MH/Unit</td><td></td><td>$20</td><td></td><td></td><td>**$20.00**</td></tr>
<tr><td>Bath Vanities in place</td><td>064110</td><td>Vanities</td><td>$1,010.00</td><td>$950.00</td><td>$1,110.00</td><td>$940.00</td><td>**$1,000.00**</td></tr>
<tr><td>Buy HM Door Frames</td><td>081213</td><td>Opening</td><td>$250.00</td><td>$235.00</td><td>$170.00</td><td>$240.00</td><td>**$215.00**</td></tr>
<tr><td>Buy Relight Glass Kits</td><td>081318</td><td>EA</td><td>$65.00</td><td>$60.00</td><td>$55.00</td><td>$60.00</td><td>**$58.33**</td></tr>
<tr><td>Install Wood Doors</td><td>081418</td><td>MH/Leaf</td><td>$1.00</td><td>$1.50</td><td>$0.85</td><td>$1.15</td><td>**$1.17**</td></tr>
<tr><td>Install Door Hardware</td><td>087100</td><td>MH/Set</td><td>$5.00</td><td>$1.00</td><td>$1.10</td><td>$1.25</td><td>**$1.12**</td></tr>
<tr><td>GWB and Metal Studs/Sub</td><td>091500</td><td>SFW</td><td>$9.00</td><td>$8.00</td><td>$5.50</td><td>$8.25</td><td>**$7.25**</td></tr>
<tr><td>Interior Painting in place</td><td>099000</td><td>SFW</td><td>$0.50</td><td>$0.50</td><td>$0.75</td><td>$2.50</td><td>**$1.25**</td></tr>
<tr><td>Buy TP Dispenser</td><td>102813</td><td>EA</td><td>$15.00</td><td>$18.00</td><td>$22.00</td><td>$20.00</td><td>**$20.00**</td></tr>
<tr><td>Install Mirrors</td><td>102800</td><td>MH/EA</td><td>$1.00</td><td>$0.50</td><td>$1.25</td><td>$0.75</td><td>**$0.83**</td></tr>
<tr><td>Purchase Window Blinds *</td><td>122000</td><td>EA</td><td>$250</td><td>$173</td><td>$225</td><td>$150</td><td>**$182.67**</td></tr>
</table>

* Note: Excludes motorized blinds on the retail project.

were motorized and cost $2,000 per window, then that would bring the average significantly up. The solution is to compute averages for all of the like-items and disqualify extremely high or low figures. Also the estimator who uses the database should scan down the types of projects to find one or two which are the most similar to the one being currently bid. A combination of these two procedures, and using some discretion to pick and choose material prices and labor productivity rates, will yield the most reliable estimate. Table 16.1 is a sample estimating database for various items which Evergreen might use when finishing or trimming out the Olympic Hotel case study.

Summary

Construction close-out requires completion of all construction work and preparing and submitting all documentation required by the contract to consider the project contractually complete. The project manager, cost engineer or cost accountant, and superintendent work closely together to ensure close-out procedures are comprehensive and efficient. Good close-out procedures typically result in higher contractor profits and satisfied clients. Physical completion of construction is the responsibility of the superintendent and management of close-out documentation and financial close-out is the responsibility of the PM and cost engineer.

As the project nears completion, the general contractor will schedule the client and design team to conduct a formal punch list walkthrough. Any deficiencies noted during this inspection are placed on the punch list for future re-inspection. All deficiencies on the formal punch list must be corrected before the contract can be closed out. A significant project milestone is achieving substantial completion, which indicates that the project can be used for its intended purpose. The architect decides when the project is substantially complete and issues a certificate of substantial completion. But the client cannot move in to their new project until a certificate of occupancy has also been issued by the City.

The general contractor's project manager, with support of his or her cost engineer, is responsible for financial and contractual close-out of the project. This involves issuing final change orders to subcontractors and major material suppliers and securing their final and unconditional lien releases. As-built drawings, operation and maintenance manuals, warranties, and test reports must be assembled. The project manager or project engineer should develop a close-out log early in the project to manage the timely submission of all close-out documents. An efficient close-out of all project activities allows the contractor to receive its final payment and release of retention, which often approximately equals its fee. An effective close-out process does not increase fee per se, but it protects the contractor's fee, as extended general conditions cost has a way of eroding fee quickly.

Development of a true as-built estimate involves combing actual man-hours and actual material costs with actual installed quantities. The cost codes which were assigned during the original estimate, and utilized throughout the cost control process, should also be applied to the as-built estimate line items. As-built estimates may be developed by the project cost engineer but should be input to the contractor's estimating database by the staff estimator. Extremely high prices and extremely low prices should be vetted so that they do not skew the average. But no project is exactly average; they all have nuances so the range of prices should be reviewed to find a project which has similar characteristics to the one which is being bid.

Review questions

1. List three close-out items a cost engineer may be tasked to expedite.
2. If close-out lasts two to three months, what happens to a contractor's fee?
3. Who on the GC's team should manage punch list completion and why him or her?
4. When does liability for LDs stop?
5. What is the AHJ looking for when they inspect for the C of O?
6. How long is a standard building warranty?
7. Why does a project owner want surplus finish materials?
8. Before a project owner releases complete retention, what do they want from the GC? Note there are several answers to this question.
9. What happens to a $5,000 warranty budget which is not used within the one year warranty period on a) a lump sum project, b) a T&M project, and/or c) a GMP project?

Exercises

1. Prepare a close-out log for the hotel case study. Include at least ten subcontractor and supplier categories. Make whatever assumptions are necessary.

2. Assume you were the PE for the construction of a new aerospace manufacturing facility. The facility manager has made a warranty claim regarding the fire alarm system – evidently it goes off everyday at 11:00 am and everyone gets to take an early lunch. You have notified the electrical subcontractor, but the subcontractor has failed to take any action. What action should you take?

3. Draw a flow chart of a material procurement process (pick one from the hotel case study) starting with the initial request for quotation and receipt of bid, and ending with receipt of the product warranty. Which of these activities would be a 'get-to' for the PE or cost engineer? You may need to refer to other chapters in the book for this exercise.

4. Our case study contractor, ECC, has received partial TCO approval from the AHJ for floors 1 and 2 and the garage. Olympic Hotel and Resort wants to conduct a 'soft' opening for a small one-week conference for 50 guests. The hotel only has 30 rooms on the second floor so a nearby hotel will handle the balance and Olympic has arranged for a shuttle service. The restaurant is not yet leased out so a caterer will be providing meals.

    a. What financial considerations will you as ECC's jobsite cost engineer need to make to keep your and the client's costs separate?
    b. What warranty issues need to be addressed?
    c. This is not a safety textbook, but what safety issues need to be addressed?
    d. What accommodations did you and the hotel need to make to convince the AHJ to issue you a TCO?

17

Time value of money

Introduction

The concept of time value of money (TVM) is basically defined as follows: *the benefit of having a stipulated sum of money today is greater than the benefit of having the same amount of money in the future.* TVM is a connection between the importance of a positive cash flow for contractors which is enhanced by fair and timely pay requests and negotiating and being paid for change orders. TVM is a popular topic in the study of engineering economics.

This is an advanced cost accounting and financial management topic and is not necessarily project-based, but more chief executive officer (CEO) and chief financial officer (CFO) related. This topic is very relevant to project managers (PMs) and cost engineers, as it is important that the jobsite team understands that 'money costs money' and if a contractor needs to borrow money from the bank to pay their monthly labor and material invoices, they will be operating with a negative cash flow, or 'in the red.' It would be unusual to find a general contractor (GC) including the cost of a construction loan in their estimate. This can be validated by looking back at the jobsite general conditions estimate included in Chapter 5 as well as the expanded blank general conditions template included on the eResource. There is not a line item for a construction loan. The contractor has no intention of financing the project for its client. If a PM operates in the red and has a negative cash flow, they will quickly be receiving a phone call or visit from the CFO. A construction application of TVM is more relevant to larger construction projects with longer durations than smaller and faster projects.

These financial terms are often used synonymously but have slight variations in definition, particularly as related to TVM, including:

- Price: the amount that is or was charged;
- Cost: the amount that was paid, or given up;
- Value: the worth an object has to someone, not necessarily the price; and
- Worth: the value that someone else is willing to pay.

One important premise of time value of money is the concept of 'equivalence.' A study of TVM makes analysis of former, current, and future investments as financially equivalent as possible. TVM does not factor non-quantitative elements such as wants and wishes and personal preferences. This chapter will discuss present value, future value, interest, inflation, and other value considerations. The cost of money is also related to our next two advanced cost engineering chapters on taxes and real estate development. An understanding of TVM helps companies and individuals when analyzing their current and future investments.

Present value

Present value (PV) is the current worth of future cash given an established rate of return and a time period, usually in years. The PV formula is the basic formula behind all aspects of time value of money. The PV formula is comprised of four variables as listed in the following and is commonly represented as:

$$PV = FV/(1+i)^n$$

- Present value (PV) is the value of an investment today, at time zero;
- Future value (FV) is the value of today's investment at a future point of time;
- 'n' is the amount of periods, often in years, in the future; and
- 'i' is the assumed interest rate an investment is projected to earn.

Any of these variables can be solved if one knows values for the other three. For example, what is the present value of $105,000 which can be garnered one year from now at an interest rate of 5%?

$$PV = \$105,000/(1 + .05)^1$$
$$PV = \$105,000/1.05 = \$100,000$$

The present value formulas make use of algorithms – a math concept that we all learned in high school but most of us do not use in everyday life. There are many financial calculators, software systems, and electronic spreadsheets that now do most of the math for us, including several web services. It is easiest for us to use pre-established tables which resolve for most of the more common combinations of years and interest. An abbreviated PV table is included as Table 17.1.

The best way to learn these formulas and concepts is through examples and exercises. If you had won the lottery and had an option to receive an established sum of money today compared to $500,000 ten years from now, assuming an interest rate of 3%, how much would be your bottom-line settlement today? Obviously $1 or $10,000 is too low. Since money today is worth more than money tomorrow, anything close to $500,000 today would be an easy decision to forego the payment in the future. The present value of this example is resolved as follows:

Table 17.1 Present value

	What is today's value of $100 'n' years in the future at an assumed 'i' interest rate?								
n	i - interest:								
Years:	1%	2%	3%	4%	5%	7%	10%	12%	15%
1	$99	$98	$97	$96	$95	$94	$91	$89	$87
2	$98	$96	$94	$93	$91	$87	$83	$80	$76
5	$95	$91	$86	$82	$78	$71	$62	$57	$50
7	$93	$87	$81	$76	$71	$62	$51	$45	$38
10	$91	$82	$74	$68	$61	$51	$39	$32	$25
12	$89	$79	$70	$63	$56	$44	$32	$26	$19
15	$86	$74	$64	$56	$48	$36	$24	$18	$12
20	$82	$67	$55	$45	$38	$26	$15	$10	$6
25	$78	$61	$48	$38	$30	$18	$9	$6	$3
40	$67	$45	$31	$21	$14	$7	$2	$1	$0.40

$$PV = \$500,000/(1 + .03)^{10}$$
$$PV = \$500,000/1.35135 = \$370,000$$
Or using Table 17.1 as a shortcut: $500,000 x 74/100 = $370,000

Any sum greater than $370,000 would result in choosing today's money versus $500,000 ten years from now. Any sum less than $370,000 would result in foregoing today's payment and taking the $500,000 ten years from now.

Future value

An expected value in the future is 'discounted' to reflect today's value, assuming a positive return on investment or inflation rate. Essentially one dollar in the future is not worth as much as one dollar today; the future dollar is worth less than today's dollar, hence discounted. The FV formula is essentially an inverse of the PV formula and is reflected by the following:

$$FV = PV(1 + i)^n$$

An abbreviated FV table is included here as Table 17.2. A simple $1 million investment made today at 5% interest in 20 years would have a value of $2,650,000 as represented in the following formula. This also can be calculated simply from our table by multiplying $1,000,000 by 265/100.

$$FV = \$1,000,000(1 + .05)^{20} = \$2,650,000$$

Table 17.2 Future value

n	i - interest:								
	What is the future value of today's $100 in 'n' years at an assumed 'i' interest rate?								
Years:	1%	2%	3%	4%	5%	7%	10%	12%	15%
1	$101	$102	$103	$104	$105	$107	$110	$112	$115
2	$102	$104	$106	$108	$110	$115	$121	$125	$132
5	$105	$110	$116	$122	$128	$140	$161	$176	$201
7	$107	$115	$123	$132	$141	$161	$195	$221	$226
10	$111	$122	$134	$148	$163	$197	$259	$311	$405
12	$113	$127	$143	$160	$180	$225	$314	$390	$535
15	$116	$135	$156	$180	$208	$276	$418	$547	$814
20	$122	$149	$181	$219	$265	$387	$673	$965	$1,637
25	$128	$164	$209	$267	$339	$543	$1,084	$1,700	$3,292
40	$149	$221	$323	$480	$704	$1,495	$4,526	$9,305	$26,786

In addition to calculating the future value of a stipulated sum, the concept of time value of money also considers FVs of an annuity. An annuity is an equal series of payments received over equal time periods. An annuity is also known as a uniform payment series. The PV and FV formulas discussed earlier are supplemented by an additional variable 'A' for annuity. For example, let's assume your great uncle left you a $1 million inheritance annuity. On the surface, an annuity paying $50,000 per year over a 20-year period appears to be worth $1 million. But actually a PV investment of only $670,000 made today at 5% interest would be equivalent to this annuity. Conversely if $1 million were invested today at 5% interest to fund this $50,000 annuity, it would last much longer than 20 years. In fact if maintenance or management costs of the investment firm holding the annuity were ignored, 5% simple interest would produce exactly $50,000 annually and the original investment would be self-perpetuating and last into infinity. There are many more advanced formulas and scenarios derived from the PV and annuity formulas that we will leave to another chapter for another day.

Interest

Interest is simply the *cost of money.* Interest explains why we have a study of time value of money. Interest is how much one party will pay another to use their money. When you put your money into a bank savings account, the bank will pay you interest, say 2%, to use your money. They then loan your money to your neighbor and charge your neighbor 4% to use 'their' money. Interest rates from banks and investors depend upon many factors including market rates, risk of the borrower, the Federal Reserve Bank (FED) interest rate, the borrower's intended use of the money, alternative investments available, inflation, and others. The higher a developer or contractor's risk factor and potential for bankruptcy as determined by the lender, the higher will be the interest rate. There are many different ways to look at interest rates including:

- Simple interest rate,
- Compounding interest rate,
- Effective interest rate,
- Nominal interest rate,
- Annual percentage rate (APR),
- Annual percentage yield (APY),
- Adjustable rate mortgage (ARM),
- Variable payment rate, and others.

Because an understanding of interest is not always 'simple,' borrowers might be confused when they thought they had signed up for a 7% home mortgage interest rate, but actually ended up paying almost 7.5% because they were being charged interest on top of interest by the bank. This is known as *compounding interest*. In 1968 Congress passed the Truth in Lending Act which requires loan institutions to also report the total annual rate that borrowers will be paying. An abbreviated interest table, comparing both simple and compounding interest, is included as Table 17.3. Note that the compounding interest Table 17.3 is very similar to the future value Table 17.2.

Another popular concept in a study of finance and economics is the *rule of 70*, also known as the *rule of 72*. It may also be technically known as the rule of 69.3, but 72 works well for many variable combinations. This rule implies that money will double when the years of an investment are multiplied by the whole numeric interest rate that the investment is earning, accounting for compounding interest and the result is 72. The rule of 72 is mathematically shown in the following equation:

$$i \times n = 72 \text{ (or 70 or 69.3)}$$

For example, money will double, or an investment will double when:

- 2% is invested for 35 or 36 years (2 x 35 = 70, or 2 x 36 = 72);
- 5% is invested for 14 years, or 14% for 5 years (5 x 14 = 70);
- 6% is invested for 12 years, or 12% for 6 years (6 x 12 = 72);
- 7% is invested for 10 years, or 10% for 7 years (7 x 10 = 70); and others.

We have introduced various financial ratios throughout this book on cost accounting and financial management. The interest rate is an important contributor to many of these ratios for an investor or developer who focuses on their return on investment (ROI), return on equity (ROE), or rate of return (ROR). Simply put, investors would forego spending their own money today for the opportunity of more money tomorrow. They like to have their 'money working for them,' essentially 'making money off of money.'

Table 17.3 Simple and compounding interest

	If $100 was invested today at 'i' interest rate, how much would it be worth in 'n' years?								
				Simple Interest					
i	*n - years:*								
interest:	*1*	*2*	*3*	*4*	*5*	*10*	*15*	*20*	*25*
1%	101	102	103	104	105	110	115	120	125
2%	102	104	106	108	110	120	130	140	150
3%	103	106	109	112	115	130	145	160	175
4%	104	108	112	116	120	140	160	180	200
5%	105	110	115	120	125	150	175	200	225
7%	107	114	121	128	135	170	205	240	275
10%	110	120	130	140	150	200	250	300	350
12%	112	124	136	148	160	220	280	340	400
15%	115	130	145	160	175	250	325	400	475
				Compounded Interest					
i	*n - years:*								
interest:	*1*	*2*	*3*	*4*	*5*	*10*	*15*	*20*	*25*
1%	101	102	103	104	105	110	116	122	128
2%	102	104	106	108	110	121	134	148	163
3%	103	106	109	113	116	134	155	180	209
4%	104	108	112	117	122	148	180	219	266
5%	105	110	116	122	128	163	208	265	338
7%	107	114	123	131	140	196	275	386	541
10%	110	121	133	146	161	259	417	672	1082
12%	112	125	140	157	176	310	546	962	1695
15%	115	132	152	175	201	404	813	1635	3289

Public clients, such as the City, borrow money from private investors in the form of bonds. These are different from bid bonds and performance and payment bonds that are part of construction estimating and contracting. If the City wants to build a new library or fire station or elementary school, or improve its roads, it will generate capital in one of a few ways:

1. Pass a bond which increases property taxes, such that the public pays for the new construction project. The taxpayers get to vote on whether they feel the investment is for the general 'public good' and they are willing to pay their share.
2. Sell bonds to private investors, essentially borrowing money from the bond holders at an agreed interest rate, say 5% for a set period of years, such as 30 years.
3. Enter into a public-private-partnership (PPP) where a private developer finances, designs, and builds the project and leases it back to the City.

Inflation

Inflation is factored into many financial studies, and although many economists focus their whole careers on making these predictions, it is not pure science as no one 'knows' exactly what the inflation rate will be next year. Inflation generally represents that *prices will increase in the future*. A consumer's purchasing power is progressively and systematically reduced by inflation. The United States Treasury and the FED work to stabilize the economy to avoid excessive spikes in interest and inflation rates. Table 17.4 reflects historical inflation rates in the United States for the past several decades. Inflation has considerably stabilized since the 1980s when we saw inflation rates exceeding 12%.

Inflation (f) variables can be substituted for interest in our present value and future value equations as shown in the following. Any of these variables can be resolved for separately the same as was discussed earlier with PV and FV. Inflation undercuts the effect of interest. If you received 2% interest on the savings account certificate of deposit your grandmother gave you for your college graduation, but inflation is 3%, you would be losing money by leaving it in the bank.

$$PV = FV/(1 + f)^n$$
$$FV = PV(1 + f)^n$$

Table 17.4 United States historical annual inflation rates

Year	Rate
1970	5.6%
1975	6.9%
1980	12.5%
1985	3.8%
1990	6.1%
1995	2.5%
2000	3.4%
2005	3.4%
2010	1.5%
2011	3.0%
2012	1.7%
2013	1.5%
2014	0.8%
2015	0.7%
2016	2.1%

Inflation is difficult to factor into construction estimates both by contractors and by project owners. If contractors include an inflation factor into a competitive lump sum bid project, they likely would not be the successful bidder as their competition would have either not included an inflation factor or included a lower one. It is commonplace for contractors to insert a clause in a negotiated contract protecting them from labor and material price escalations and passing that risk on to project owners. Project owners do not receive inflation exclusion clauses very well and an attempt by a contractor to change order for price increases as a result of inflation often ends up as a claim or project dispute. Contract clauses such as 'adjusted for inflation' are common risk mitigation methods for buyers and sellers in all aspects of life, not just construction.

Banks will offer a borrower a fixed 30-year mortgage with a 5% interest rate such that the home owner will have exactly the same payment (with a different mix of principle and interest amounts) over the life of the loan. But an adjustable rate mortgage may be offered as well at 4%, which can be 'adjusted for inflation' which allows the bank to slowly increase the interest, often with a cap such as 2% above the initial rate.

Other value considerations

There are many value considerations in a study of cost accounting and financial management relevant to project managers, cost engineers, and CFOs. This chapter has focused on time value of money, specifically present value and future value. In this section we briefly introduce a few additional value considerations, many of which also connect with previous chapters in this book covering cash flow, pay requests, and change orders and upcoming advanced chapters on taxes and pro forma.

Economy of scale is similar to an economic study of diminishing marginal returns; the first scoop of ice cream was great, but the fourth not necessarily so. For example, a small boutique hotel which has 30 rooms on two floors costs considerably more per hotel room to build than a five-story hotel with 120 rooms. If the fixed costs of an elevator, restaurant, and swimming pool were factored in, and both the two-story and five-story hotels had these same facilities, this would amplify the construction unit price cost if figured on the count of hotel rooms. The construction estimator needs to consider economies of scale when applying unit pricing to concrete and doors and other direct work items. It costs much more per cubic yard (CY) to place 10 CY of concrete in one day versus 100 CY. This easily applies to quantity of garage stalls as well; the larger the quantities, the more economical are the unit prices.

At some point a larger hotel plan would require an additional restaurant and an additional elevator, or in the case of a parking garage, an additional underground floor. These major additions are stepped cost increases. This is a similar concept to fixed and variable overhead costs discussed in Chapter 6. An investor must consider *sunk costs* when adding money to an initial investment. When the original investment is 'sunk,' it is gone and it cannot be recovered. Additional investments may enhance the investor's ability to recover the initial investment, but are not guaranteed to do so. An investor, including a real estate developer, may be faced with the dilemma of walking away from their original investment to avoid the old adage of 'pouring good money after bad.'

Another popular financial concept is the *cost recovery period*, or *payback*. An investor will put money into a building, such as improved mechanical equipment or new windows or insulation, if resulting savings in energy and utility costs will pay back that investment in a given period of time, often seven years. If it takes longer to pay back the investment, then the developer often passes on the increased construction cost. If it is less than or quicker than seven years, then the developer may choose to invest in capital improvements today.

As stated, contractors do not typically factor a construction loan or interest on a loan in their estimates. But this may not be the case with subcontractors. As discussed in Chapter 13, 'Cash flow' and Chapter 14, 'Payment requests,' subcontractors are *paid after the GC is paid* and may be out-of-pocket two months before they are paid; and third-tier subcontractors even longer. Essentially subcontractors are used as a means for temporary construction financing by the project owner. This is why it is important that general contractors check subcontractors' financial strengths when they are evaluating awarding a contract to a 'best-value' subcontractor and not necessarily a low bid. An understanding of time value of money also re-enforces a contractor's interest in *front-loading* their pay request schedule of values so they are operating 'in the black' and using the owner's money and not their own.

Developers and contractors often prepare early budget estimates to assist with a variety of financial decisions. If design documents are preliminary, they may rely on facility dollars per square foot of floor ($/SF) *database pricing* published in sources such as RS Means. But although there are thousands of prices in any typical published database, they are national averages and are not specific to any one project or location. For example, if Evergreen Construction Company (ECC) or their client had used a published database they may have developed a budget for our case study project of:

Hotel:	100,000 SF @ $140/SF =	$ 14,000,000
Garage:	52,000 SF @ $60/SF =	$ 3,120,000
Total Project:		$ 17,120,000

But our case study project does not exactly fit the national average; no project really ever is 'average.' ECC's contract price is $24.5 million which is considerably higher than this budget. It was to be built approximately 100 miles north and west of Olympia, Washington so a location modifier of 1.05 (or an additional 5%) needs to be added just for Olympia and even more for a remote suburban area. The average database sizes for a hotel and garage are based on 50,000 SF and 160,000 SF respectively, but our case study project is different and therefore additional size modifiers need to be applied. This is because of an economy of scale as discussed earlier. If another database was to have been used, either an older one (published prices would be lower as explained in present value and future value earlier) or a future database edition (prices would be higher) then an additional TVM modifier would be necessary as well. Incorporation of all of these modifiers would adjust the budget to slightly greater than $20 million and closer to ECC's contract value.

Summary

Although a thorough knowledge of time value of money is typically relevant to a contractor's CEO and CFO, it is also important for a jobsite project manager and cost engineer to have a basic understanding. Money costs money, and if a project owner makes a late payment or negotiations are slow on change orders, or if there is an unfavorable amount of retention held, this will affect a contractor's cash flow. Most contractors do not include short-term construction loans in their estimates.

The concept of time value of money is simple: $100 yesterday was worth more than it is today, and today's $100 will be worth more than it will be tomorrow. There are many formulas related to present value, future value, annuities, interest, and inflation. Most of the combinations have already been figured out for the TVM novice and are available on websites.

Interest is the cost of money, or the amount one party (the bank) will pay another (a bank depositor) to use their money. One is the borrower and the other the lender. Use of money is not free. Inflation measures the historical changes of values. If milk cost $5.00 per gallon in one year, and $6.00 in another year, that $1.00 increase reflects a 20% price increase, or 20% inflation.

There are virtually an unlimited amount of other value considerations we could have introduced in this chapter. One value consideration which specifically relates to time value of money, especially for a construction estimator, is adjustments made to published database prices from national averages to reflect a specific project's location, size, and the year it is being built. The study of TVM is of particular relevance to investors and real estate developers as will be reflected in the last chapter of this book.

Review questions

1. What are some other combinations of interest and years that meet the rule of 72?
2. Why does the combination of seven years and 10% not equal 200 in the simple interest Table 17.3?
3. Why does a GC concern itself with a subcontractor's financial strength?
4. The construction contract between ECC and Northwest Resorts is for $24.5 million. Is this the 'cost,' 'value,' 'price,' or 'worth'? Note: it could be more than one of these choices.
5. Why should a PM and cost engineer be aware of TVM?
6. At what interest rate (i) would a PV of $100 be the same as a FV of $100 in any given year (n)?
7. When would a PV cost figure be greater than a FV cost figure?

Exercises

1. Do you have money in a savings account? You should! How much interest are you earning? Is it enough? If left in place at this interest rate, when would your savings account double?
2. If your great aunt had left you an inheritance of $10,000 in 1980, and it had been locked up in a certificate of deposit at an interest rate equal to the inflation rate at that time, i.e. the same amount per year (lucky you!), how much would it be worth today at a) a simple interest rate, and/or b) a compounding interest rate?
3. What is the FV of a $1 million investment made today ten years from now with an anticipated 15% annual ROE?
4. What is the present value of a lump sum $300,000 inheritance received 40 years from now assuming a 2% inflation rate?
5. If an investment is made today of $400,000 at a guaranteed rate of 5%, how many years will it need to be invested to realize a FV of $1,000,000?

6. Assume that your $250,000 investment made today will realize you $2,500,000 in 20 years in the future. What was your annual ROE?

7. Using the Olympic Hotel and Resort estimate from Chapter 4 as your baseline, and adjustment factors from a current copy of RS Means, or other published estimating database, how much would the hotel cost if it were built in:
   a. the same location during the same time but at one half the size?
   b. the same location during the same time but at twice the size?
   c. the same location in 2012 at the same size?
   d. in Seattle, Washington in 2018 at the same size?
   e. in Little Rock, Arkansas in 2010 at the same size?
   f. in Little Rock, Arkansas in 2018 at the same size?
   g. in Little Rock, Arkansas in 2010 at one half the size?
   h. in Little Rock, Arkansas in 2018 at twice the size?

18

Taxes and audits

Introduction

Approximately one third of a construction company's net profits are paid to the United States Government and state and local governments as income taxes. Additional taxes are also paid to various government agencies in the form of business or excise tax, payroll or labor tax, sales tax, as well as others. This book has been focused on the project manager (PM) and the jobsite cost accountant's management of financial issues that are project related. Taxes are, per se, not necessarily project related but more a corporate issue and a focus of the chief financial and chief executive officers (CFO and CEO). But it is important for the jobsite team to know how their efforts affect the company's bottom line – that is, net profit after tax. This chapter will introduce income and other taxes and how those taxes affect financial ratios. Taxes are a necessary part of everyone's lives. Taxes finance the government and support the needs of society, such as roads and schools.

This chapter is not about how to prepare an income tax form for the Internal Revenue Service (IRS), rather those types of tax issues are best left to other textbooks which focus on taxes and accounting for general business purposes. This chapter introduces income taxes and some of the 'costs' or deductions that contractors will take in order to reduce or postpone taxes, which improves their immediate cash flow position. The focus is also on the revenue associated with a company's construction business and not that from separate or independent investments of the equity partners. Tax laws are complicated and are always changing, and tax laws from one state are not necessarily the same as another. Contractors will consult with tax accountants, certified public accountants (CPA), and tax attorneys for these types of advanced business issues, similar to consulting with attorneys for contract preparation and claim resolutions. But again, these business exchanges are made at the corporate level and not the jobsite level.

The amount an employee receives on their pay check is much less than what it costs an employer to employ that individual. There are several add-ons to the cost of labor, whether that be indirect labor including the project manager and superintendent, or direct craft labor such as carpenters and laborers. All of those add-ons are grouped together as labor burden, which includes both labor benefits and labor taxes. Labor taxes are also influenced by various Federal and state agencies.

Audits are an advanced financial management topic and Federal income tax audits involve the IRS. Several types of audits are introduced in this chapter as well, and many of them involve the jobsite project team including monthly and end-of-project audits from the client. Many larger contractors will also perform in-house audits of their own operations and several efficiency aspects of individual projects. Audits and taxes are definitely advanced project management topics and are more home office and CEO and CFO related than the project management and cost engineering coverage in previous chapters.

Income tax: Federal, state, and local

Income taxes are paid by all individuals and corporations, including construction companies. Basically the more an entity makes in income, the more tax they have to pay. Taxes can exceed one third of income, which is substantial. There is likely not one other single cost code or account a construction company experiences which amounts to the total taxes it pays, therefore income taxes receive a lot of attention from construction accountants and the CFO. Construction companies are project-based; that is one of the several reasons construction is different from other industries such as food service or airplane manufacturing. Each construction project needs to produce sufficient fee to cover home office overhead and leave the company with profit, enough profit that they can pay their taxes. The revenue equations presented throughout this book have included:

Revenue = Cost + Overhead and profit (or Fee)
Cost = Direct construction cost + Jobsite indirect construction cost
Direct construction cost = Labor + Material + Equipment + Subcontractors
Profit = Revenue – Expenses, including direct and indirect jobsite costs
Net profit = Fee – Home office overhead
After tax profits, or Pure profit = Net profit – Taxes

On the jobsite, the superintendent and project manager are excited that they just completed a $50,000,000 elementary school on time and under budget with acceptable quality and no safety accidents. But they did not make $50,000,000 for their company. They produced a significant amount of revenue which gets allocated to a variety of sources as depicted earlier, but the corporate officers and owners of the construction firm are interested in the final answer, after tax profits. This is the true gauge of whether the company was successful and it is this figure that is then re-invested into the company or distributed to the equity owners as dividends.

Planning for, avoiding, reducing, or deferring taxes is acceptable, if done legally, but tax fraud is illegal and results in substantial fines and potentially imprisonment. Contractors will attempt to reduce their tax obligation by taking whatever deductions are allowed according to the tax laws. Some of the largest deductions to a contractor's tax obligations, beyond the jobsite direct and indirect cost of construction, include:

- Cost of home office overhead, including office rent;
- Salaries and bonuses and dividends paid to corporate officers;
- Equipment and real estate depreciation;
- Charitable contributions; and others.

Different tax structures for different types of organizations

As described in Chapter 2, construction companies can be organized in a variety of fashions, and each of those different types of legal organizational structures is regarded uniquely by the IRS with respect to Federal income tax. In addition, the method a contractor chooses to report revenue, whether that is cash, accrual, percentage complete, or completed contract all have different tax implications.

Sole proprietorship: There is not any difference in tax applications between a sole proprietorship and an individual. Companies organized as sole proprietorships have their business profits blended with the individual's other sources of income and all are taxed at the same level. The company's income and deductions are reported on IRS Schedule C which is attached to the standard 1040 tax form.

Partnership: A partnership may be comprised of two or several individuals. The company is not taxed, rather the partners share the company's income, according to their respective equity shares, which is reported on their individual 1040 tax forms and taxed the same as if it was a sole proprietorship. The partnership is transparent with respect to income and taxes and personal liability. If there is not sufficient equity in the partnership to pay for a liability, the creditors can access the general partners' personal assets.

Corporation: A corporation is significantly different from a sole proprietorship or partnership in that the corporation is a unique and separate entity, not a person. An individual, or group of individuals, can 'incorporate' and contribute equity into this separate entity. Unlike partnerships, the assets of the individuals are protected and separated from liabilities of the corporation. Investors' exposure to loss is limited to their equity contributions. The corporation is taxed and if there is remaining profit which is not re-invested in the corporation, it is distributed as dividends to the equity investors. The investors then report dividends on their individual 1040 tax forms (and Schedule D) and pay 15% or 20% tax on the distribution, depending on their individual tax bracket. Corporations are therefore subject to double tax, first at the corporate level as business income and then at the individual level as dividends. Corporations are also known as 'C' corporations.

'S' corporation: An S corporation (or S Corp) is a subset of C corporations. They are not subject to the double tax that a standard corporation pays; all of the profits flow direct to the investors and are combined with their other personal incomes. The personal liability of the equity investor is protected similar to corporations. S Corps are also known as closely held corporations and their quantity of shareholders is limited. Many larger corporations start out as S Corps.

Limited liability partnerships (LLPs): LLPs are not used for contractors but some of the other built environment participants such as architects or consultants may be organized as LLPs. Each individual partner is protected from the acts of their other partners. For tax purposes LLPs are very similar to corporations.

Limited liability corporations (LLCs): LLCs are a subset of LLPs and also have many tax implications similar to standard corporations. Personal assets are protected from corporate taxes.

Many real estate developers will set up a separate LLC for each construction project; that way if one project fails it does not affect the value of the other projects – each is stand-alone. See Chapter 19 for additional discussion on real estate development. Different states have different rules with respect to LLPs and LLCs. In some states LLCs pay an additional franchise tax.

Joint ventures (JVs): A JV is a partnership usually created for just one project. This is common when two contractors will form a JV to bid a very large construction project as each alone does not have sufficient bonding capacity. General contractors (GCs) and architects may also form a JV for a design-build project. Once that project is completed, the JV will dissolve. The JV is similar to other partnerships with respect to taxes.

Strategic alliances: These are not separate companies but rather just informal agreements that two or more firms will work together to potentially capture a new project or market share. For example, a general contractor may form a strategic alliance with large mechanical and electrical subcontractors to work together in their bid preparation on a specific project. There is no tax consideration with these informal arrangements.

Federal income tax rates

The IRS levies personal income tax on a 'marginal' basis. Taxes are a percentage of net income; lower levels of income pay a lesser percentage of tax. As the income levels rise, the individual jumps into the next marginal tax level or tax bracket. In 2017 Congress passed major tax reform legislation which modified the tax brackets, standard deductions, itemized deductions, and incorporated other tax law changes. The personal and married filing jointly categories changed slightly. In both cases there were previously and still are seven different graduated tax brackets. The tax percentage was reduced moderately for low to mid-level incomes but the greatest benefit was for those individuals with incomes over $415,000. Personal income tax levels are represented in Table 18.1 as of 2018.

If an individual makes $100,000, they do not pay 24% tax on the total earned, rather they pay a marginal tax rate of 10% on the first $9,525, and 12% on the next $29,175 ($38,700–$9,525), and 22% on the next $43,800 ($82,500–$38,700), and finally 24% on only the last $17,500 ($100,000–$82,500). This would result in a total tax obligation of:

Table 18.1 Personal income tax

	Single Filer	Married Joint Filers
Rate	Incomes	Incomes
10%	$0 to $9,525	$0 to $19,050
12%	$9,526 to $38,700	$19,051 to $77,400
22%	$38,701 to $82,500	$77,401 to $165,000
24%	$82,501 to $157,500	$165,001 to $315,000
32%	$157,501 to $200,000	$315,001 to $400,000
35%	$200,001 to $500,000	$401,000 to $600,000
37%	Over $500,000	Over $600,000

$$\$ 9,525 @ 10\% = \$ 952$$
$$\$29,175 @ 12\% = \$ 3,501$$
$$\$43,800 @ 22\% = \$ 9,636$$
$$\$17,500 @ 24\% = \$ 4,200 \text{ (24\% marginal tax bracket)}$$

Total earned = $100,000 and total tax = $18,289
The average or effective tax for that individual is therefore $18,289/$100,000 = 18%

Utilizing the previous tax brackets from 2017 yielded a total tax obligation of $21,037, at a maximum bracket of 28%, which was 21% average tax. This results in a total tax savings from 2017 to 2018 for this individual of $2,748 or 13% at the $100,000 income level. This ignores other tax changes made for standard and itemized deductions as those could have numerous combinations and are specific for any individual.

In 2017 there were eight different tax brackets for corporations with graduated rates ranging from 15% for income less than $50,000 to 38% and 39% brackets for those companies with very high incomes. The 2018 tax changes enacted by Congress were much more significant for corporate than personal incomes. Now all corporations, regardless of their profit level, pay a flat 21% tax which is almost one half of what was charged previously for most companies. Corporations are no longer taxed 'marginally.' A corporation earning $100,000 in 2018 will pay 21% on that amount or a total of $21,000, which is a higher total than the individual tax obligation at that level as shown earlier. But the individual was in the 24% marginal bracket and any additional income beyond $100,000 would be taxed at that level and the total will quickly approach and exceed the new 21% tax rates of corporations. Conversely a corporation in 2017 earning $100,000 would have had their total and average taxes computed as follows:

$$\$50,000 @ 15\% = \$7,500$$
$$\$25,000 @ 25\% = \$6,250$$
$$\$25,000 @ 34\% = \$8,500 \text{ (34\% marginal tax bracket)}$$

Total earned = $100,000 and total tax = $22,250
$22,250/$100,000 = 22% average tax rate

Individual tax rates increase to 37% at $500,000, which are therefore much higher and almost double corporate tax rates. The decisions of how a construction company is organized, whether they be a sole proprietorship or partnership or S corporation or other, is based on a variety of reasons, including potential tax liabilities. The new tax bill also includes tax deductions for these pass-through business entities to make them more compatible with C corporations. The new 2018 tax laws may have an effect on how contractors legally organize themselves going forward. Additional IRS rules and guidance are available at www.irs.gov.

Other tax implications

Capital gains taxes are levied against profits from the sale of stocks, real estate, or construction equipment held longer than one year. Capital gains tax rates are 15% or 20% for individuals depending on the tax bracket, and are at a lower percentage than corporate income taxes. Profits from sales of personal assets held less than a year are taxed at the normal income levels. Capital gains for corporations are taxed as ordinary income. Losses from sales can be proactively reduced from the prior two years' income or carried forward for up to 20 years to offset future income.

Dividends paid to equity partners are reported on their personal 1040 tax forms and taxed at 15% or 20% depending on their individual tax bracket, similar to capital gains. Dividends are reported on the corporate income statement as business expenditures.

Estimated taxes: Contractors and other businesses do not make one large payment to the IRS on April 15 of each year. The Federal Government does not want contractors to be using what the government considers as their money. Contractors and other businesses must pay-as-they-go and make quarterly estimated income tax payments. These occur on April 15, June 15, September 15, and January 15 of the following year.

Net after tax ratios: Chapter 7 introduced financial ratios that were derived from the contractor's balance sheet and income statement. Many of those ratios can be figured before-taxes as well as after-taxes. Some of those which are related to taxes include return on equity, return on investment, return on assets, and rate of return.

State and local income taxes: Most states have separate income taxes with rates which range from approximately 3% in North Dakota to over 13% in California (as of 2017). Separate annual tax forms must be filed to states and localities from Federal tax forms. Both individuals and corporations have traditionally been able to reduce their total Federal income tax owed by the amount they paid to the states, but this has been reduced in the 2018 tax laws. Only seven states **do not** have income tax, which are: Alaska, Florida, Nevada, South Dakota, Texas, Washington, and Wyoming.

Business, property, and sales taxes

There are a variety of other taxes that both individuals and corporations are exposed to. Every state is different; there are not any generalizations with respect to these other taxes and the tax rules are always changing. If a contractor is proposing to bid on a project which is out of state or out of their normal area of business, they should learn all they can about that particular area's tax rules. Here are a few examples of different types of taxes in states with different tax rules. Note that these cannot be 'added-up' as some try to do, especially those planning on a place to start a business or retire from business. There are many variables which should be applied by a business or investment tax professional before decisions are formalized.

Some states and cities will have additional corporate taxes that are revenue-based and not necessarily income-based. These taxes are known as *business tax* or *business and occupational tax* or *excise tax*. These terms are used differently in different locations so a good comparison is difficult to make. Some states levy these taxes just on alcohol and cigarettes, others on property, and others on the total revenue a company brings in, regardless of expenditures or profits. Washington State, combined with local jurisdictions, charges approximately 1% state excise tax against total revenue, regardless of expenditures, on all elements of the built environment including contractors, designers, and consultants.

Property taxes, if applicable, are usually paid by the client or developer to cities, counties, and states, after the completion of the project, and annually thereafter, based upon the assessed value of the improved property. All states charge property taxes in various manners, some on just the improvements and some on the entire value, including the lot. New Jersey has the highest effective property tax rate at 2.38% with Hawaii and others as low as 0.3% as of 2017, but county and city rates may apply as well.

Sales taxes are a separate tax and may be assigned to individual project values or just some portion of construction cost such as a sales tax on materials but not on labor. In some cases taxes are excluded from bids and contracts but are collected from the client by the general contractor with monthly pay requests and paid to the state. There are only five states currently **without** sales taxes, which are: Alaska, Delaware, Montana, New Hampshire, and Oregon. All other 45 states have sales taxes and they range from a low of about 4% in Hawaii to over 10% in Louisiana including county and city taxes. Alaska does not have a state sales tax but some local jurisdictions there have sales taxes.

Labor taxes

Direct labor costs contractors much more than just the wages the craftsmen and administrative supervisors receive on their pay check. Contractors pay an additional markup, or percentage add-on on top of all of the wages they pay; this is known as labor burden. Labor burden is not a fee or profit markup, but it is a direct cost of doing work. The amount of this markup is not established at the jobsite level by project managers or superintendents. Rather labor burden is determined at the CFO and CEO level. The burden is usually a journal-entry charge to the jobsite and is not accompanied with a separate invoice. Labor burden has two major components, labor taxes and labor benefits, as reflected in the following equation:

Labor burden = Labor taxes + Labor benefits

Required labor taxes

Some will refer to all of labor burden as either labor taxes or labor benefits, but they are distinct and have different costs and rates for different types of labor. Labor taxes are also known as payroll taxes. Labor taxes are government determined and have at least four major elements which contractors are required to pay; this includes:

* *Social security*, also known as FICA, was created by the Federal Insurance Compensation Act. The employer pays one half or approximately 6.2% up to the first $120,000 of wages (changes yearly) for FICA contributions and the employee pays the other half as a withholding from their weekly check, for a total tax of 12.4%.

- *Medicare* is also a joint contribution from the employer and the employee similar to FICA and amounts to approximately 1.45% each for a total of 2.9% on the first $117,000 earned.
- *Unemployment tax* has two elements, Federal and state:
  ◦ Federal unemployment tax costs 6% on the first $7,000 of income, but much less if there is a state unemployment tax; and
  ◦ State unemployment tax varies by state. The amount of unemployment tax percentage paid by any company is proportional to the amount of unemployment claims they experience from personnel lay-offs.
- *Workers' compensation insurance* markups, or *workers' comp*, vary considerably due to a variety of factors including the safety record of the contractor and its associated experience modification rate (EMR), the potential safety risk of the labor craft, and differences between indirect and direct labor. The baseline EMR is 1.0. Contractors which have a higher incident rate of safety accidents have an EMR rate greater than 1.0 and those with fewer accidents a rate below 1.0. Some crafts are more prone to accidents and they will have a higher workers compensation rate. Indirect salaried employees have a much lower chance of a safety accident and therefore have a much lower workers compensation rate than a direct work craftsman.

Labor taxes are the same percentage markup for union and merit shop contractors, but they are wage dependent. The higher the wage of the craftsman, the higher generally will be the labor taxes.

Labor benefits

Labor benefits are also known as fringe benefits and are determined by the contractor and include a variety of items. These are not 'taxes' per se but are voluntary contributions. This discussion is included here as they are a significant portion of the total labor burden that is attached to wages. If contractors are signatory to labor unions they will likely have higher labor benefits than do contractors which employ merit shop labor. Some items which may be included with labor benefits include:

- Health insurance
- Dental insurance
- Eye insurance
- Disability insurance
- Life insurance
- Union dues
- Pensions and retirement
- Use of company cars and cell phones
- Vacation
- Sick leave
- Bonus
- Education and training
- Safety add-on
- Supervision add-on
- Small tools add-on

Some labor benefits, such as medical insurance, are a shared cost between the employee and the employer. Depending upon contract terms and the definition of reimbursable costs on open-book projects, some contractors may include more or less of these potential labor benefits in their labor burden rates.

Labor burden

There are significantly different labor burden rates for direct craft labor (carpenter and electrician) than from indirect labor (project manager and superintendent and CEO and CFO). In addition, different crafts or trades have different rates depending on the type of work and associated safety risk; ironworkers and electricians are subject to more safety incidents than painters and landscapers. In addition, there are a variety of union issues which affect labor burden, such as some trades do not provide their own tools, the contractor does. Subcontractors are responsible for paying their own labor burden; the general contractor does not get involved with subcontractor burden. Table 18.2 includes several trades' wage rates and their associated labor taxes and labor burdens.

It would be very cumbersome for a construction company CFO to invoice or journal entry different jobsite labor burden rates for each type of craft or administrative labor category. Most contractors will develop a 'blended' burden rate at the beginning of the year which is based on labor mixes from the prior year. A blended burden rate for a general contractor which is signatory to the carpenters, laborers, ironworkers, and cement finishers unions might be 55% or higher. Contractors which utilize merit shop labor will have a lower total labor burden percentage markup due to a mix of fewer labor benefits which were listed earlier, maybe 25%. Usually direct and indirect labor is kept separate as the burden rate of indirect labor is so much less than direct (maybe 30%), but even these may be blended on certain projects, again depending on contract terms. In addition some contractors will invoice their clients on open-book projects a 'loaded' wage rate which includes the base rate plus a blended burden rate. Contractors may choose to do this as a loaded wage reflects the total cost of an hour of direct labor, but it is difficult to substantiate during an audit of an open-book project as discussed later. The loaded wage rate for laborers in Table 18.2 would therefore be $51 per hour. Some contractors include other markups with labor burden such as liability insurance, but these are volume-dependent and not labor-dependent and would not be accurate on projects with a different mix of direct labor versus subcontracted labor.

Construction projects which receive Federal funding require contractors to pay direct craft employees a 'prevailing wage rate,' which is the wage rate most common to the area being worked plus labor taxes. This is also known as the Davis-Bacon wage rate. The prevailing wage rate for carpenters on the Olympic Hotel case study project in Table 18.2 would therefore be their base wage plus the combined labor burden. The construction market in this area is predominantly

Table 18.2 Labor burden

Construction Craft	Base Wage	Labor Taxes	Labor Benefits	Total Burden
Carpenters	$39	$8	$12	$20
Cement Masons	$32	$6	$10	$16
Electricians	$43	$9	$13	$22
Equipment Operators	$38	$10	$11	$21
Ironworkers	$41	$12	$18	$30
Laborers	$33	$8	$10	$18
Plumbers	$46	$11	$16	$27

union and therefore the union wage rate is 'prevailing.' It is up to the contractor whether they want to contribute additional labor benefits beyond the prevailing wage rates.

Labor burden is charged on a contractor's own direct and indirect labor only. Labor burden is not added to material costs, equipment rental, or subcontractors. Subcontractors are required to factor their own labor burden within their bid prices to the general contractor.

Audits

When one hears they are going to be 'audited' it often causes a shiver to go down their spine. But it doesn't have to be that way. There are various reasons for audits and various types of audits. They are not necessarily an attempt to find errors or wrong-doing, audits are often just a system to verify that what the individual or corporation is doing is correct. An audit can be looked upon as a system of 'checks and balances.' Audits may be performed or prepared for a variety of stakeholders including:

- Equity partners or stockholders,
- Contractor's bank or lending institution,
- Contractor's surety or bonding agency,
- Clients,
- Internal Revenue Service,
- National Labor Relations Board (NLRB) and others.

Internal audits

Contractors may perform audits on themselves. These are also known as internal or management audits and are conducted for a variety of reasons. One is to test that their systems of quality and safety and schedule and cost controls are meeting corporate objectives. Contractors may survey clients and others outside of the company to verify that the company is meeting the needs of the industry. Sometimes the contractor, or project team, may be too close to a situation to see it clearly and a separate independent test or inspection may be fruitful. Construction company corporate officers may also perform scheduled or unscheduled audits of jobsites to verify their field management teams are performing as expected. Another internal audit generated by the jobsite team is a 'lessons-learned' audit which for some is the last phase of a construction project, after close-out. This is an internal look at the project and an assessment of what worked and what didn't and what can be done better next time. Project managers and superintendents also evaluate the performance of their subcontractors during this post-project audit and share that with their internal colleagues.

Another reason any corporation, construction company or others, may perform internal audits is to validate that its financial statements will support outside scrutiny. Because equity partners and the bank must be assured that the financial data reported by the construction company is accurate, most contractors, especially mid- to larger-sized contractors, will first perform an independent internal audit of their financial conditions with their own accountants and bookkeepers, and then employ the services of an outside certified public accountant to perform an audit. The better prepared are the books by the contractor's own accountants, the easier it is for an outside

CPA to perform an independent audit which validates the financial results or findings. This will also reduce the cost of the CPA's audit as they customarily work on an hourly basis. The goal of the CPA audit is for the auditor to prepare a report and provide an official 'unqualified opinion, without exception' that the financial statements of the construction company are accurate and have been prepared and presented fairly. In addition to checking the balance sheet and income statement, CPAs will review a contractor's backlog, their accounts receivable and payable, project manager monthly fee forecasts, as well as investigate for any potential internal thefts or fraud committed by the contractor's own employees.

As discussed throughout this book, the construction industry is unique from others, and a financial audit of a construction company is also unique and complicated for a variety of reasons including the long-term nature of construction contracts, billings on a percent complete basis, client withholding of retention, allocation of equipment and overhead to specific projects, and several others. Certified public accountants are by nature 'independent' accountants. Although the construction company pays them for their services, they are required to have taken and passed the Uniform CPA Examination and abide by the guidelines established by the American Institute of Certified Public Accountants. Contractors may also utilize the services of a CPA to assist with their tax preparation, financial planning, and internal stock value determination.

Client audits

Depending upon the terms of the construction contract, clients may have the ability to audit the contractor's books for that specific project, especially if it is a private client on a negotiated open-book project. These types of audits may be performed each month or only once at the completion of the project and part of the final financial close-out. Regardless of when they are conducted, the contract must clearly state the owner's audit requirements and the contractor must be given sufficient notice to prepare their books for review. The general contractor's jobsite team will play a major role in client audits, especially the project manager and the jobsite cost accountant.

Monthly audits are customarily performed in conjunction with the monthly pay request. It would be difficult for the project owner's auditor or accountant to validate the costs of each pay request before it is paid, so usually the audit is performed after the pay request has been processed. In this case the February pay request for Evergreen Construction Company on the Olympic Hotel and Resort project is processed as of February 28 and Northwest Resorts pays them on or about March 10. The amount to be paid is that which was agreed to during the jobsite walkthrough conducted on February 25 with the owner and architect and bank. Then after March 10, the client's auditor, which may be an outside accounting firm, will review the contractor's actual cost records for the month of February to determine if the recorded and projected costs were fair. Some of the items the auditor will review include:

- Job cost history report;
- Subcontractor invoices and lien releases and payments made to subcontractors;
- Labor report, including time sheets and validation of checks paid;
- Material report and payments made to suppliers;
- Equipment ledger; and others if requested.

The goal of the auditor is to validate that all of the charges which have been attributed to his or her client's project were actually spent on this project. For example, assume this were a $5,000,000 grocery store project yet the contractor had paid a swimming pool subcontractor $100,000, or the general contractor charged the project for new tires for the contractor's CEO's luxury sedan, or if a subcontractor was paid more than what had been agreed in their contract. These would all be reasons for the auditor to raise a red flag. These of course are extreme exceptions. Sometimes audits simply discover journal entry or cost coding errors. Minor variances may occur and the contractor will simply make an adjustment on their next monthly pay request. But the process to prepare for the auditor and respond to their inquiries on a monthly basis may be burdensome and expensive for the GC's project manager and needs to be accounted for in their jobsite general conditions estimate, often with the addition of a jobsite cost accountant on the team.

More common than the monthly audit is an end of the project audit. All of the same types of documents are reviewed, including validation of percentage add-ons or journal entries to the job cost such as insurance and taxes and labor burdens. This audit may take several weeks to perform and may involve more than one accountant. Again the auditor will prepare a final report, present it to his or her client, who will then discuss exceptions with the contractor. The final release of retention is often predicated upon a successful audit report. This type of audit may also be necessary for an owner's representative who must report to a board of directors or other stakeholders within the client's organization to assure that the owner's representative managed their finances appropriately.

There are many *other types of audits* that the contractor's home office or jobsite team will experience. The IRS may audit the contractor's tax statements. If prevailing wages are paid on federally funded projects, the NLRB will check the 'certified payroll' to see that appropriate wages were paid to the craftsmen including reviewing time sheets and pay check stubs. Subcontractors on a prevailing wage project are subjected to the same requirements. If a contractor utilizes union craftsmen, the unions may audit the contractor on an annual basis to verify that sufficient labor benefits were paid for pensions and training. Contractors and projects will also experience a variety of compliance audits or inspections from public agencies such as the City for code compliance and the Occupational Safety and Health Administration (OSHA) for safety violations.

Summary

Taxes are an important part of our personal and business lives and they cannot be avoided. A good understanding of taxes and appropriate tax planning can be beneficial to all businesses, especially contractors. The net profit that a contractor makes is after deduction of all of its jobsite and home office expenditures from revenue, as well as a deduction for income taxes. The resultant amount or percentage is surprisingly small, especially compared to the risks associated with the construction business.

Individual Federal income taxes are levied on a marginal level. As income levels rise, the percentage of tax that must be paid also steps up to the next marginal level. The highest level of tax is considered the marginal tax rate. The total tax paid, divided into the reported income, is the average or effective tax. Individuals and corporations have different tax levels and there are tax advantages of operating as a proprietor or corporation at different levels of income.

There are many other state taxes beyond the Federal income tax that individuals and companies must pay including income taxes, business taxes, property taxes, and sales taxes. It is important for a contractor pursuing a project in a new state to completely understand the local tax implications before submitting a bid. Labor taxes are required by the government and include social security and worker's compensation. Additional labor benefits may also be added to the cost of direct and indirect wages. The total labor burden markup includes labor taxes plus labor benefits which can exceed 50% of the cost of wages.

An audit is a check on the accuracy of the financial affairs of the contractor. Audits are performed for a variety of internal reasons including cost performance of individuals and the team and reporting to company equity investors. External audits are performed by the client in an open-book project and those requirements must be understood in the contract before commencing work. They can be quite arduous if the contractor's books are not kept consistent and accurate throughout the course of construction. Other groups may perform external audits including the IRS for tax purposes and unions to verify labor benefits have been paid.

Review questions

1. Which craft in Table 18.2 has the highest: a) bare wage rate, b) total labor burden, c) percentage markup for labor burden, and/or d) loaded wage rate?
2. What is the difference between a marginal tax rate and an effective or average tax rate?
3. Which person costs the contractor a higher percentage of labor burden, the cost accountant or the carpenter foreman?
4. What is the difference between labor burden and labor benefits?
5. Other than roads and schools, what might taxes be used to pay for?
6. What is the difference between a loaded wage rate and a blended wage rate?
7. What types of legal business organizational structures are subject to double income tax?

Exercises

1. What are some of the different ways a construction firm owner, who is both an equity partner and a company employee, can financially benefit through the ownership of a construction company? There are several.
2. Utilizing our Federal income tax table in this chapter, at a pre-tax net income of $500,000, which has the highest *total tax* liability, an individual or a C corporation?
3. Utilizing our Federal income tax table in this chapter, at a pre-tax net income of $20,000,000, which has the highest *marginal tax rate*, an individual or a C corporation?
4. Assuming five craftsmen for each of the crafts in Table 18.2, a) what would be the total blended labor burden percentage markup, b) what would happen to that rate if you eliminated the craft with the highest burden and doubled the manpower from the craft with the lowest burden, and c) what would happen to the original rate if you eliminated the lowest burdened craft and doubled the manpower from the highest craft?
5. If contractors charge a blended labor burden rate to a client, and alter their craft mix as discussed in Exercise Four, what would that potentially do to their profit?

6. If a contractor charges the blended labor burden rate you calculated from Exercise Four, (a), to a client, but applied it to both direct and indirect labor, how would this affect their profit?
7. Every state was not included in all of the tax lists. Where does your state stand amongst the highest and the lowest discussed?
8. Why do you think the State of Alaska does not have any sales or state income taxes?
9. Not discussed in this book, but what might be a downside of a GC forming a strategic alliance with an architect or a mechanical subcontractor?

19

Developer's pro forma

Introduction

Construction is only one aspect of the real estate development business and it is actually one of the most black and white, easily understood, and most manageable elements. There are many more risky aspects to development than construction and there are many steps the developer goes through long before the contractor places the first concrete into the footing forms. This final chapter is going to diverge away from a total contractor-focus and discuss the role of the real estate developer and its relation to construction cost and accounting methods. This is not a textbook on real estate development; there are several other good books on that topic, many of which are available from the Urban Land Institute (ULI, reference uli.org). The National Association of Industrial and Office Parks is the commercial real estate developer's association and is also a valuable resource for someone interested in the development arena of the built environment (reference naiop.org).

Real estate development and the developer's pro forma are advanced topics but a brief overview is relevant to the construction project manager (PM) who has contracted with a developer. This chapter first discusses the business of real estate development and many of the steps or phases of the development process. Financial aspects of real estate development which may have an effect on the role of the construction PM include preparation of the developer's pro forma, construction loan applications, monthly draws and payments, and additional financial ratios which govern development decisions. The real estate arena has many additional terms and phrases – those introduced here are also included in the glossary.

Real estate development as a business

Many participants in the built environment (BE) industries envy the role of the real estate developer. The developer is thought to be an extreme risk taker and the individual or company which realizes the largest profits. Very few new construction management graduates will start their careers in real estate development; typically they need to obtain ten years or so of experience in some facet of the BE and an understanding of the role of the developer. Some of the more experienced BE participants who aspire to one day make that transition and become developers themselves include:

- Attorneys, especially land-use attorneys,
- Architects,
- Constructor lenders,
- General contractors (GCs), including construction project managers,
- Subcontractors,
- Real estate brokers, and
- Consultants.

Not all clients or project owners are developers. Many owners 'build-to-suit' projects for just their specific needs and personal occupancy and do not plan on selling the project when completed. These owners will likely establish a separate ownership structure for the building from the business; this separation will provide the real estate investment with tax and asset protection advantages. In that regard these project owners may encounter many of the development steps including pro forma preparation and loan applications as discussed here, but because they are also the tenant, their projects are much less risky from a lender's perspective.

There are several different forms a real estate development company will take when operating solely for the purpose of improving property for financial gain. Individuals, such as this author, are sole-proprietor developers. Two or more individuals can join to comprise a development partnership and as they grow and hire more employees, they become a development company. Land developers purchase bare land and improve it 'horizontally' by subdividing and installing roads and utilities. This is also common with speculative home development. Vertical developers build office buildings and apartments and lease the building and either quickly 'flip' it and sell it or hold it long term. Developers may also purchase an existing structure and perform renovations or change the use and also 'flip' it and realize a short-term profit. Development agents or fee developers manage the entire process for an investor without putting any of their own money or equity into the project. All of these forms of real estate development are grouped together in this chapter which will simply use the term 'developer.'

The reason many of the built environment participants listed earlier want to enter into the development field is because they view development as extremely profitable. Development can be profitable, but it is also very risky and often prone to bankruptcy. As with any business, the more risks there are, the higher are the expected profits. Development success relies on many factors including the experience of the developer, market timing, and sometimes just luck. For example, an apartment building may begin the planning process during a boom rental market but takes five years to complete including design and permitting and construction. The building is finally ready to lease but now the market has been saturated and the apartments cannot be rented at the rates that had been expected when the pro forma was developed. It is very difficult to predict market timing and the risk associated with missing it demands high profits. One way a developer will limit their risk is to create different ownership structures for each property. Resorts International, Inc. (RII) is the client for the Olympic Hotel and Resort sample case study project in this book, but actually RII will not own any of its projects directly, rather they will establish a separate and independent limited liability corporation (LLC) for each one. That way if one project was to have financial difficulty, it would not affect that of the other properties; essentially the original owners or investors in RII would be shielded from creditors on any one individual project. In this case, Northwest Resorts, LLC has been set up as the owner of the Olympic Hotel project.

Some general contractors will also operate as developers. This is particularly true with speculative home builders which will develop property horizontally, build the houses with their own crews or subcontractors, and market and sell their finished product. Other commercial GCs may establish a development arm or division which would also be a separate and independent company, often a limited liability corporation, with the officers of the company as equity owners. This developer will then hire itself as a contractor to build projects, essentially creating work for itself. In this instance the two entities will actually enter into a construction contract and attempt to maintain an arms-length relationship. This author's first-hand knowledge reflects that this is easier said than done.

The development process

As stated, there are many early and lengthy steps the developer goes through that most contractors are not aware of. In the case of a bid project, the contractor comes on board once all of this front-end work has been completed, including having financing and permits all in place. Some of the built environment participants who want to gain access to the development business are unaware of all of these steps and the amount of time and financial resources a developer must have in order to bring a project to the construction phase. This section describes these phases and some of the activities which occur during each phase as well as additional BE participants the developer will employ through the development process.

There is much more to the development process than purchasing a bare piece of land and drafting drawings and constructing a building. Market research analysis up front and marketing the product throughout the process are extremely important. Management of the entire process requires leadership and creativity and is the *synergy* which often results in a successful endeavor. The development process is represented by the following real estate development formula:

Land + Design + Financing + Construction + Marketing + Synergy = Development

Development phases

Real Estate Development, Principles and Process, from Urban Land Institute (ULI), describes eight stages or phases of any real estate development project. Other resources describe similar phases. These eight phases are:

1. Inception of an idea including market research of the product and location;
2. Refinement of the idea including site selection;

3. Feasibility study including preliminary design and additional market analysis. This includes creation of a due-diligence study and refinement of the developer's pro forma;
4. Contract negotiation with the property seller, permitting agencies, contractors, lenders, and potentially tenants and buyers. Many of these contracts will be 'contingent upon' completion of the feasibility study and successful negotiation of the other contracts;
5. Contract execution including property purchase and construction contracts and securing financing;
6. Construction;
7. Construction completion including close-out and occupancy; and
8. Operations.

The design of the project progresses from development stage one all the way through stage five and potentially into stage six as is the case with design-build projects and some negotiated fast-track projects. The design phases most projects experience include:

*   Programming, development phases 1 and 2;
*   Conceptual design, phases 2 and 3;
*   Schematic design, phase 3;
*   Design development, phase 4; and
*   Construction documents, phase 5.

During development phase 6, construction, the design may continue to progress with final finish selections by the design team and submittal and shop drawing preparation from the construction team.

Development team

The three major participants in any built environment project include the owner, designer, and contractor. Their relationship and construction team organizations were described in Chapter 3. The developer's organization is much larger and is represented as a very horizontal organization chart in that most of these participants work for the developer direct. Architectural and construction organizations are more vertical, with personnel and subcontractors and consultants working for and reporting through other entities. Other than the scope of the general contractor, which is considered a hard cost, most of the work products from these development team participants are considered soft costs. They are necessary elements but you often cannot see the direct result in the final building. Some of these firms are employed very early in the development process, as early as phases 1 through 3, and before lender financing has been secured. The developer must therefore pay for these services out of pocket, or from owner's equity (OE). Some of the companies and individuals employed early as part of the due-diligence team are listed here. These and other members of the developer's organization are shown graphically in Figure 19.1:

*   Architect,
*   Surveyor,
*   Geotechnical engineer,

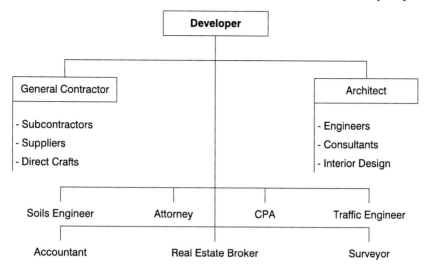

Figure 19.1 Developer's organization chart

- Real estate broker,
- Land use attorney,
- Accountant, and others.

Pro forma

The developer's pro forma is a complicated equation with many variables and often customized by different developers for different projects. The pro forma can be thought of globally as the financial analysis necessary to determine whether the project 'performs' and is a viable real estate investment. The bank requires the pro forma to be included with the loan application package. Internal development equity partners and potential external investors also rely on the pro forma as a decision-making tool. A very rough 'back-of-the-envelope' pro forma is developed early in the process, as early as development phases 1 or 2 described earlier. The pro forma continues to be refined and modified all the way through the development process and even through construction completion (phase 7) and into occupancy (phase 8). The pro forma combines many cost estimates and compares the estimated cost of the project to the expected value. If the projected value is 10–15% greater than the estimated cost, then the project is likely a 'go' whereas if the expected costs approach or exceed value, then the project is a 'no-go.' This concept is also known to developers as 'does it pencil?' Or does the real estate development investment make sense? Either way, the developer may go back into the pro forma and change one or more of the variable values which may provide a different result. All real estate development costs in the pro forma are grouped into four major categories:

- Land purchase;
- Land improvement, infrastructure, or horizontal costs;
- Soft costs, including architectural design and permits and financing; and
- Hard costs, primarily general construction cost.

Evergreen Development, LLC

1234 West First Street Mixed Use Development Project

Abreviated Pro Forma

Hard Costs:	Quantity	Units	Unit Price		Cost
Land Purchase	50,000	SFL	50 $/SFL		$2,500,000
Site Improvements	30,000	SFL	12 $/SFL		$360,000
Height Limit/Floors	60	Ft. @	12 ft/flr =	5 Floors	
Building:					
Square Footage	20,000	Footprint	5 Floors =	100,000 GSF	
Shell	100,000	GSF	120 $/GSF		$12,000,000
TI	100,000	GSF	65 $/GSF		$6,500,000
Parking Garage	0	GSF	30 $/GSF		$0
Subtotal Construction:					$18,500,000
Subtotal Hard Costs:					$21,000,000
Soft Costs:					
Design Fees			6% of Constr		$1,110,000
Developer Overhead/PM			3% of Constr		$555,000
Property Taxes During Construction			1.25% of Land		$31,250
Marketing/Broker	1	LS			$135,000
Construction Loan Int.	six months of		5.0% of 3/4 of total cost		$480,100
Permits			1% of Constr		$185,000
3rd Party Inspections			1% of Constr		$185,000
Accounting	1	LS			$40,000
Legal	1	LS			$45,000
Subtotal Soft Costs:					$2,766,350
Developer Fee			5% of SC & HC		$1,188,318
Subtotal Soft Costs:					$3,954,668
Subtotal Hard & Soft Costs:					$24,954,668
Contingency:			5% of Cost		$1,247,733
Total Costs:					**$26,202,401**
Value:					
Net Leasable Area	100,000	GSF	80% of GSF =	80,000 NSF	
Gross Revenue	80,000	NSF	30 $/NSF		$2,400,000
Garage Efficiency	0	SF	350 SF/Stall =	0 Stalls	
Garage Revenue	0	Stalls	200 $/Stall/Mo		$0
Subtotal Revenue:					$2,400,000
Less Vacancy			8%		$192,000
Less Operation Costs:					
Property Tax			1.2% of Cost		$314,429
Utilities			1.1 $/GSF		$110,000
Repairs & Maint.			1.2 $/GSF		$120,000
Prop. Management			3.0% of Rev		$72,000
Subtotal Operation Costs					$616,429
Net Revenue, or Net Operating Income (NOI)					**$1,591,571**
Capitalization Rate			6%		
Value:			NOI/Cap Rate =		**$26,526,186**

If Value > Costs +10-15%, then GO, if not, then NO GO

Figure 19.2 Pro forma

The developer may need to finance the first three of these out of owner's equity and may only receive a construction loan on the construction cost, timing and value dependent. A sample abbreviated pro forma for a spec office building case study prepared by Evergreen Development, LLC is included as Figure 19.2. A detailed pro forma could be over 100 pages long with estimate spreadsheets backing up each of the line items shown here. Evergreen Development is a development arm of Evergreen Construction Company comprised of all of its officers including the chief executive and chief financial officers (CEO and CFO). Evergreen Development always contracts with Evergreen Construction for new development projects. A live Excel version of the pro forma is included on the eResource. Note that the general construction estimate is shown as only one line item in the pro forma. This supports the point made earlier that construction is only one element in the entire process and the developer's risk and early cash outflow is much greater than just construction. The importance of the pro forma is included in this discussion of financial management as construction project managers, estimators, and project accountants all need to know where they fit into the developer's financial equation.

One of the most important pro forma variables is the *capitalization rate*, or 'cap rate.' The cap rate is used to determine the estimated economic value of the building upon completion, from which a loan amount can be calculated. Some banks have standard cap rates they use when evaluating a construction loan. The method to calculate the cap rate is:

$$(\text{Loan interest} \times \text{percentage loaned}) + (\text{ROR} \times \text{percentage invested}), \text{ or}$$
$$(5\% \times 75\%) + (10\% \times 25\%) = 6.0\%$$

The higher the expected value of the building the more the developer can borrow, possibly using proceeds from a loan on one project to leverage another. Leverage and loans are discussed later. The cap rate is mathematically divided into the anticipated net operating income (NOI) to produce the expected value as follows:

$$\text{NOI} / \text{Cap rate} = \text{Value, or}$$
$$\$1,591,571 / 6.0\% = \$26,526,186$$

Construction loan

Loans and interest and financing and loan fees are lengthy topics worthy of their own chapter if not their own textbook, which is the case with many general financial studies in business schools. This book has revolved around construction jobsite cost accounting and in this chapter the contractor's client, the developer, is added to that coverage. A brief discussion of contractor financing is included in this section along with a more lengthy analysis of the developer's construction loan and securing permanent financing. This chapter briefly introduces many additional financial terms for construction project managers and jobsite cost engineers. A more detailed exploration will be reserved for future corporate CEOs and CFOs.

Contractor financing

As a former senior project manager and a construction company owner, this author will tell you that commercial and custom home construction PMs do not finance construction projects. Contractors are not in the business of providing short-term loans to their clients. It is the client's responsibility to finance the construction project. This is evidenced by peeking back at the case study construction estimate and noting that there is not a line item in the estimate for a construction loan or debt financing. Contractors strive to operate in the black and use creative methods to generate positive, not negative, cash flows, as has been described throughout this book. The two primary types of construction company loans or financing occasionally needed and managed by the contractor's CFO include both long- and short-term loans.

Long-term loans will be obtained from commercial banks by contractors for expansion of business operations which may involve purchase of land or buildings or construction equipment. Long-term loans are generally for a payback period in excess of one year. A long-term loan is reflected as both an asset (cash) and a liability (owed to bank) on the contractor's balance sheet. Payments on a long-term loan will be a combination of interest and principal. The ratio of interest to principle is much larger (almost all interest) early in the loan period and much smaller (almost all principle) late in the life of the loan as shown in Figure 19.3. Buildings will be financed for a term of 15–30 years and equipment often five years. As an alternative to obtaining long-term expansion financing from a bank, the contractor's ownership can contribute additional equity, known as owner's equity, or take on additional partners or sell stock, which dilutes ownership. OE typically requires a higher rate of return (ROR) as that money is directly at risk, so if a contractor can avoid increasing their OE they will do so.

An unsecured loan means the bank is not requiring any assets or collateral be placed in escrow or guaranteed and made available in case the contractor defaults on its loan. A secured loan means the bank attaches itself to some asset made available by the borrower. In a sense, the bank becomes a part-owner or a first lien holder on the property. Lenders obviously prefer secured over

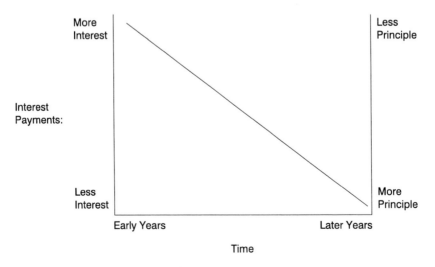

Figure 19.3 Interest to principle ratio

unsecured loans. Construction firm ownership will be required to pledge their personal assets, such as their homes or other real estate property, to be used as collateral guaranteeing repayment of an equipment or real estate loan to the bank.

Short-term loans may be necessary to fund day-to-day construction activities if the construction projects themselves cannot be self-funding. Short-term loans are expected to be paid back in one year or less. Often only interest is paid on the loan, sometimes all up front, and the principal is paid in full as a balloon payment when the loan has matured. If the interest for a short-term loan is paid in full up front this is known as 'discounting' the loan. Short-term loans are set up as 'lines of credit' where the contractor has a pre-approved amount, say $500,000, which is at the bank and available whenever needed. A small fee such as 0.5–1% will be paid to the bank to keep this money in place and, when borrowed, will be subject to current interest rates until the funds are paid back. Contractors may also use additional subcontractors and fewer direct craftsmen as a means of providing short-term financing and minimizing their need to borrow against their line of credit. Subcontractors are not 'paid until the GC is paid' as discussed previously with cash flow and pay request chapters.

Development financing

The developer's banker is one of his or her most important partners and team members. Just as contractors need bonding sureties, insurance companies, attorneys, and accountants as strategic partners, developers rely on banks to fund their projects and enhance their leverage. There are many aspects of development financing that relate to construction accounting and financial management. The following sections are going to describe leverage, permanent financing, and construction financing. In construction as well as in other industries, stakeholders talk in terms of 'where is the money?' Without the bank, no one gets paid. The following formula represents how far away is the 'money' from the subcontractors, suppliers, and craftsmen:

> $ Bank > Developer > GC > Subcontractors > Suppliers and craftsmen $

Real estate developers possibly understand the importance of *leverage* more than in any other industry. Leverage means to get as much mileage out of one's equity as possible. The developer tries to use other people's money to make money for themselves. The best way to express this is through the use of examples:

> Example One: This developer has $1,000,000 in cash. He buys a commercial building for $800,000 and puts $200,000 worth of improvements in it, all out of his own pocket. At the completion of the project he sells the building for $1,200,000. This represents a $200,000 profit, or 20%, which most of us would feel is a substantial return on our investment (ROI).

Example Two: This developer also has $1,000,000 in cash. He buys five buildings each costing $800,000 but only puts 25% down with his own money or $200,000 for each building and borrows the balance of $600,000 for each building from the bank. He now has $1,000,000 of his own money in these projects and has borrowed $3,000,000. He then borrows an additional $1,000,000 against his equity in the buildings and puts $200,000 worth of improvements in each project. This developer eventually also sells each of his five improved properties for $1,200,000 for a total of $6,000,000. He owes the bank a total of $4,000,000 for the two loans. This leaves him with $2,000,000. Based on his original investment of $1,000,000 he has realized a $1,000,000 profit, or 100% ROI. This is leverage. Note that the interest cost of short-term loans was omitted from this scenario for simplicity. Developers would usually pay for the loan from the loan proceeds itself, just as they would pay for building materials such as carpet.

Connected with the concept of leverage is the 'risk-reward' concept. This was introduced prior when a contractor would gauge the amount of fee it would add to a construction estimate based upon several risk factors perceived. In the case of real estate development, the more a developer extends itself and borrows from others, the more it is leveraged out, the greater are the risks assumed and therefore the greater are the expected rates of return. This risk-reward relationship is shown in the following formulas:

$$> \text{Equity}, < \text{Debt}, < \text{Risk}, < \text{ROR}$$
$$< \text{Equity}, > \text{Debt}, > \text{Risk}, > \text{ROR}$$

The developer will use money from the construction loan before he or she uses the *permanent loan* so it seems the discussion of the construction loan should be placed first. But it is presented here in reverse, because the loan application for the permanent loan happens first and a construction loan will only be secured after the permanent loan has been pre-approved. This occurs in phase 5 of the development process. Sources of permanent financing include:

- Real estate investment trusts (REITs);
- Pension funds such as the teacher's retirement fund;
- Life insurance companies;
- Commercial banks;
- Savings and loan institutions (S&Ls), also known as thrift institutions;
- Owner's or developer's equity;
- Original land owner or seller;
- Venture capitalists or limited partners;
- Commercial mortgage backed securities (CMBS); and
- Government agencies such as Veterans Affairs (VA) and Federal Housing Administration (FHA).

Developers will pay up to a 2% loan fee to a mortgage broker who will scour the market for the best permanent loan package. This fee is often a line item in the pro forma. Both permanent and construction lenders are conservative by nature. They are not in the business of owning and operating buildings, let alone incomplete construction projects in case of developer or contractor failure. The 2008 financial crisis was due in part to bank loans made to weak developers and inadequately managed construction projects as well as loans which did not require a down payment. Today it is customary for the lender to require the developer to have 25% equity in the project, or plenty of 'skin in the game.' Developers may take on silent partners or investors to contribute equity as well.

The permanent loan, also known as a mortgage loan, is a long-term loan, from 15 to 30 years. Repayment of the loan is a combination of interest (more up front) and principal (more in the later years) similar to long-term loans described earlier for contractors. The amount that will be financed is determined upon the estimated value of the completed project and the expected revenue stream or rent. The developer has to 'pitch' and 'sell' their proposed idea to the bank. The application package for the permanent loan includes many documents such as:

- Developer's history, portfolio, and references;
- Business plan;
- Developer's financial statements including balance sheet;
- Last three years of tax returns;
- Designer's resume and financial strength;
- Preliminary design documents;
- Feasibility study and pro forma;
- General contractor's resume and financial strength and possibly similar from the major subcontractors;
- General contractor's estimate and schedule;
- Signed lease agreements with tenants; and
- Building permits.

Construction loans are typically provided by commercial banks. The construction loan is a temporary short-term loan typically lasting a year or less, or until the construction project is completed. The bank may also be the same provider of the permanent loan. The construction lender will only approve a loan to a developer if it already has secured a permanent loan. The construction lender will require the design to have been completed, permits in hand, a firm construction price, an executed contract with the general contractor, and likely performance and payment bonds provided by the GC. Considering all of the items included in the permanent loan package, what this means is that the whole development deal has to come together at the same time, which is somewhat a 'chicken or egg' situation – which comes first?

During the course of construction the developer pays only interest on the construction loan, and the loan principle itself is paid in full at the end of the job, often direct from the permanent lender. This interest is also known as the 'debt service' and is a line item in the pro forma. During the course of construction the developer gathers invoices from all of its team members, including the general contractor and architect and consultants. The interest paid on the construction loan and other fees and soft costs are added in. The developer's pay request will look similar to the GC's pay request to the client. The developer's pay request is often termed a 'draw' in that they are drawing down or against the construction loan. The bank may hire an inspector who will participate in the GC's draft schedule of values review at the jobsite which occurs about the 25th of each month. The developer relies on a cooperative attitude from the GC with respect to the bank to assist with pay request approval. The GC invoices the developer at the end of the month, on or about the 30th, and the bank will usually process and pay by the 10th of the following month. This then allows the GC to process payments to its subcontractors as discussed in previous chapters.

Once construction is complete the developer will transfer out of the construction loan into a permanent loan which is typically one to two percentage points less in interest, which is a significant motivation factor to get the project complete. This permanent loan then is considered a 'take-out loan' in that the construction lender is being taken out of the equation. Many of the close-out documents discussed in Chapter 16 which were required of the general contractor by the client will also be required of the developer by the bank, including certificates of substantial completion and occupancy as well as final and unconditional lien releases.

There are many other *financial terms* and fees affecting real estate development such as: points (fee for a loan), gap loans, bridge loans, mezzanine loans, land lease or ground lease, Libor, balloon payments, adjustable rate mortgages (ARM), basis points (portion of 1%), closing costs, amortization, escrow fees, title insurance, appraisal cost, and several others. These will be left to a more advanced future financial discussion, possibly when the reader has become a construction CEO or CFO or a real estate developer.

Investment ratios

This study on cost accounting and financial management has introduced and described many financial ratios for the construction contractor, which were derived from the contractor's balance sheet and income statement. There are additional financial ratios and calculations which a developer will prepare when deciding the feasibility of a real estate development project and submitting for their loan application. The mathematical result of these ratios is ideally above or below a certain number or within some desired percentage range. Some of those ratios and financial analysis include:

- Cap rate: NOI/Property value
- DCR: Debt coverage ratio, measures loan risk = NOI/Annual debt service, >1.5
- Estimated property value = NOI/Capitalization (Cap) rate
- IDS: Income to debt service or debt service ratio,
  Annual projected income divided by loan payment, ideally less than 1.3
- IRR: Internal rate of return
- LTV: Loan to value,
  Loan amount divided by expected value, ideally 75% or less
- MC: Mortgage constant ratio = Annual debt service/Loan amount
- NOI: Net operating income = Annual income − Annual expenses
- ROA: Return on assets,
  Net profit after taxes divided by total assets, ideally 5% or greater
- ROE: Pre-tax ROE = net profit before taxes divided by equity,
  After-tax ROE = net profit after taxes divided by equity,

Greater than 5% is expected, ideally approximately 15%,
Also known as 'cash on cash'
- ROI: Return on investment = NOI/Equity investment
- ROR: Rate of return

Summary

Real estate development is a much more involved and complicated element of the built environment industries than is just construction and that is why it was reserved for the last chapter of this cost accounting and financial management book. Developers are often the contractor's clients. Development ownership can take on many forms, from individuals to large corporations. Development is thought to be a very profitable business and that is why many others involved in the BE want to try their hand at real estate development. Development can be profitable, but it is also very risky and relies on experience and relationships the developer has with many of its team members, including the bank. There is much more to real estate development than just buying land and erecting a structure. It takes a thorough understanding and research of the market and the ability to market the product along with managerial expertise. What makes this all work, especially on successful projects, is the developer's *synergy* in bringing many elements and participants together to create value. In real estate development, synergy is the 'glue' which validates that 'the whole is greater than the sum of its parts.'

There are many documents the developer must submit to the bank to obtain financing, both permanent and for construction. The document that has the most variables and is scrutinized heavily is the pro forma. The pro forma calculates the anticipated value of the project based upon land purchase costs, land improvement, hard cost of construction, and several soft costs. Contractors will secure long-term financing for business expansion and equipment purchases and maintain a line of credit to manage day-to-day expenses, but they are not in the business of providing construction financing. The developer first obtains a commitment for permanent financing and with that in hand obtains the construction loan. There are many items required by the bank for the loan application including statements of financial condition and tenant lease agreements.

Review questions

1. Looking at the leverage Example Two included earlier, how many projects could be leveraged and what ROI would a developer realize if he or she only had to put down 10% for each project?
2. What is the structure of the payback difference for a GC for a short-term versus a long-term loan?
3. If a GC is running short on operational cash, why doesn't the business owner just contribute more of his or her own personal cash?
4. What is a developer's 'draw'?
5. What is a 'debt service'?
6. What is the difference between a cap rate and an interest rate?
7. Why would a contractor keep its development operations separate from its construction operations?

8. What is the difference between a spec real estate developer and a build-to-suit client?
9. Why do so many participants in the BE want to become developers?
10. Define 'synergy.'
11. Why would a bank not want to loan a developer to purchase bare land or loan for soft costs necessary to perform a feasibility study?
12. Why does a construction loan cost 1–2% more in interest than a permanent loan?
13. Which firm would likely realize the largest profit potential, an at-risk developer or a fee developer?
14. Which firm would go bankrupt if a project was completed but lost its tenant and sat vacant for several years: a) fee developer, b) at-risk developer, c) architect, d) real estate broker, or e) GC?
15. The word 'pro forma' sounds similar to the word _____.
16. Other than their personal homes, what might equity owners be required to pledge for loan collateral?
17. The developer's pro forma continues to be modified and improved with increased accuracy throughout the development process. What other contractor-generated financial document and cost reporting process discussed earlier in this book improved accuracy throughout the course of construction?
18. Is the project contemplated in Figure 19.2 a 'go' or a 'no-go'?

Exercises

1. Putting on your construction hat, what are three items which would be included under 'land improvements'?
2. What risks does a developer have by modifying the pro forma to make the project appear more profitable than it might actually be?
3. There are many companies and individuals which might participate in a real estate development organization chart. The generic development organization chart included here has a few of them. List three more.
4. If one floor of garage below grade is added to the example pro forma at 20,000 gross square feet (GSF) using the parking stall efficiency figure of 350 SF/stall and the rental of $200/stall, what did that do to the original pro forma?
5. We introduced a lot of new terms and abbreviations and acronyms and slang in this last chapter particular to real estate development. We did not include a separate glossary specific to this chapter, but the budding real estate development student might be interested in putting one together.
6. There are literally hundreds of exercise combinations which can be created by modifying variables in Figure 19.2 pro forma. Other than the garage elements which are included in Exercise Four, choose three variables to modify by 25% each to make this a 'go' and as profitable an endeavor as possible, and/or choose another three to make it least feasible and definitely a 'no-go.'

Glossary

Accelerated depreciation: the IRS allows businesses to expense larger losses in paper values, such as for buildings and equipment, during the early useful life and smaller values during the later useful life; this helps with reducing income taxes

Accountant: individual who performs financial tasks including accounts payable and accounts receivable; reports to or may be the same as the chief financial officer; see also *bookkeeping*

Accounting cycle: a series of numerous steps beginning with a construction estimate and transitioning through buyout and construction including payments received from the client and payments made to vendors and culminating in an as-built estimate which feeds back into the company database and supports the next project estimate

Accrual accounting method: revenue is recognized when clients are invoiced and expenses are recognized when invoices are received, but not necessarily paid, usually reserved for smaller contractors

Active safety control: process that anticipates and prevents safety problems rather than just responding to and correcting issues after an accident

Activity-based costing: process of applying home office and also jobsite indirect costs to departments and projects and direct construction activities if possible

Agency construction management delivery method: a delivery method in which the client has three contracts: one with the architect, one with the general contractor, and one with the construction manager. The construction manager acts as the client's agent but has no contractual authority over the architect or the general contractor

Agreement: a document that sets forth the provisions, responsibilities, and the obligations of parties to a contract; standard forms of agreement are available from professional organizations such as the American Institute of Architects, ConsensusDocs, and others

American Institute of Architects: a national association that promotes the practice of architecture and publishes many standard contract forms used in the construction industry

As-built drawings: contractor-corrected construction drawings depicting actual dimensions, elevations, and conditions of in-place constructed work

As-built estimate: assessment in which actual costs incurred are applied to the quantities installed to develop actual unit prices and productivity rates

As-built schedule: marked-up, detailed schedule depicting actual start and completion dates, durations, deliveries, and restraint activities

Asset: physical property or items of value including real estate, buildings, equipment, furniture, and tools

Associated General Contractors of America: a national trade association primarily made up of construction firms and construction industry professionals

Audit: verify what is reported, such as construction costs in the case of an open-book negotiated project, is actual; often by a third party such as a CPA or the IRS

B&O tax: business and occupational tax collected by states and localities based on total revenue, also known as state excise tax

Back charge: general contractor charge against a subcontractor for work the general contractor performed on behalf of the subcontractor

Balance sheet: list of assets and liabilities and owner's equity such that the sum total of all assets equals the total of all liabilities plus owner's equity

Bid shopping: unethical general contractor activity of sharing subcontractor bid values with the subcontractor's competitors in order to drive down prices

Bookkeeping: generic term for all types of accounting processes and personnel

Breakeven analysis: minimum fee necessary to cover home office overhead before profit is realized

Builder's risk insurance: protects the contractor in the event that the project is damaged or destroyed while under construction

Building information models or modeling: computer design software involving multi-discipline three-dimension overlays improving constructability and reducing change orders

Build-operate-transfer delivery method: a delivery method in which a single contractor is responsible for financing the design and construction of a project and is paid an annual fee to operate the completed project for a period of time, such as 30 years

Buyout: the process of awarding subcontracts and issuing purchase orders for materials and equipment

Buyout log: a project management document that is used for planning and tracking the buyout process; usually comparing subcontractor bid values against adjusted and contracted values

C corporation: legal organizational format of a company which includes stock holders and a board of directors, often associated with larger and mature companies; also known as C corp

Cash accounting method: utilized by smaller contractors where expenses are recognized when checks are cut and revenue is recognized when a check is received, but not necessarily an exchange of hard currency

Cash flow curve: a plot of the estimated value of work to be completed each month during the construction of a project

Certificate of occupancy: a certificate or document issued by the city or municipality indicating that the completed project has been inspected and meets all code requirements

Certificate of substantial completion: a certificate signed by the client, architect, and contractor indicating the date that substantial completion was achieved

Certified public accountant: accountant qualified to perform independent financial audits and often prepares corporate statements including tax returns; requires testing and certification

Change order: modifications to contract documents made after contract award that incorporate changes in scope and adjustments in contract price and/or time. A commonly used form is AIA document G701

Change order proposal: a request for a change order submitted to the client by the contractor, or a proposed change sent to the contractor by the client requesting pricing data

Change order proposal log: a listing of all change order proposals indicating dates of initiation, approval, and incorporation as final change orders

Chief executive officer: individual at the top of a company's organizational chart; may also be the president or chairman of the board or *officer-in-charge*

Chief financial officer: individual responsible for all internal accounting and bookkeeping in an organization, usually an officer and/or equity owner in the company; may also be referred to as the *financial officer* or *financial manager*

Claim: an unresolved or post-project request for a change order

Close-out: the process of finishing all construction and paperwork required to complete the project and close-out the contract; also includes *financial close-out* and construction close-out

Close-out log: a list of all close-out tasks that is used to manage project close-out

Commissioning: a process of testing and assuring that all equipment and operating building systems are working properly, especially MEP systems

Completed contract accounting method: usually associated with speculative home builders or smaller remodeling general contractors or subcontractors where revenue and expenses are recognized when the project has been finished

Conceptual cost estimate: cost estimates developed using incomplete project documentation; also known as a schematic estimate or budget estimate or *rough-order-of-magnitude estimate*

Conditional lien release: document representing the amount of money which has been invoiced and is due to a contractor or supplier. When that money is received, the contractor or supplier will release all rights to lien the physical property, but only for the exact amount of money received

ConsensusDocs®: family of contract documents which has taken the place of the AGC contract documents

Constructability analysis: an evaluation of preferred and alternative materials and construction methods; also a design quality control process

Construction change directive: a directive or document issued by the client to the contractor to proceed with the described change order

Construction loan: short-term loan issued to a developer, usually from a commercial bank, which concludes when transferred into a permanent or take-out loan after construction is complete

Construction manager agency delivery method: see *agency construction management delivery method*

Construction manager-at-risk delivery method: a delivery method in which the client has two contracts: one with the architect and one with the construction manager/general contractor. The general contractor usually is hired early in the design process to perform preconstruction services. Once the design is completed, the construction manager/general contractor constructs the project; also known as CM/GC

Construction manager/general contractor delivery method: see *construction manager-at-risk delivery method*

Construction Specifications Institute: the professional organization that developed the original 16-division MasterFormat that is used to organize the technical specifications; today's CSI includes 49 divisions

Contract: a legally enforceable agreement between two parties

Corrected estimate: estimate that is adjusted based on subcontractor and supplier buyout costs

Cost codes: codes established in the firm's accounting system that are used for recording specific types of costs

Cost control cycle: systematic financial expense control process engaged by a contractor starting with a construction estimate progressing through buyout, cost recording, process modifications, and culminating in an as-built estimate which feeds back into the company database and supports the next project estimate

Cost engineer: project engineer reporting to a project manager whose focus is only on cost related issues such as buyout, purchasing, cost coding, cost control, work packages and others; also known as jobsite cost accountant, see also *project engineer*

Cost estimating: process of preparing the best educated anticipated final cost of a project given the parameters available

Cost loaded schedule: a schedule or spreadsheet in which the value of each activity is distributed across the activity, and monthly costs are summed to produce a cash flow curve

Cost-plus contract: a contract in which the contractor is reimbursed for stipulated direct and indirect costs associated with the construction of a project and is paid a fee to cover profit and company overhead; also known as a *time and materials contract*

Cost-plus contract with guaranteed maximum price: a cost-plus contract in which the contractor agrees to bear any construction costs that exceed the guaranteed maximum price unless the project scope of work is increased

Cost plus fixed fee contract: a cost-plus contract in which the contractor is guaranteed a fixed fee irrespective of the actual construction costs

Cost plus percentage fee contract: a cost-plus contract in which the contractor's fee is a percentage of the actual construction costs; see also *time and materials contract*

Cost recovery period: time it takes to return the cost of an investment made in the form of profits received

Craftspeople: non-managerial field labor force who construct the work, such as carpenters and electricians; also known as craftsmen or tradesmen

Critical path: the sequence of activities on a construction schedule that determine the overall project duration

Current asset: short-term item of value such as cash and accounts receivable or ability to turn an item of value into cash within approximately one year or less

Current liability: short-term debt such as accounts payable to craftsmen, material suppliers, and subcontractors usually due within one year or less

Daily job diary: a daily report prepared by the superintendent that documents important daily events including weather, visitors, work activities, deliveries, and any problems; also known as daily journal or daily report

Davis-Bacon wage rates: prevailing wage rates determined by the U.S. Department of Labor that must be met or exceeded by contractors and subcontractors on federally funded construction projects

Demobilization: physically shutting down the construction project and removing the site camp and all construction equipment

Depreciation: the loss of value of an asset such as a piece of construction equipment or building due to normal wear and tear which can be deducted from other profits to reduce income taxes; also equipment and/or real estate depreciation

Design-build delivery method: a delivery method in which the client hires a single construction company which both designs and constructs the project

Design, build, operate delivery method: a delivery method in which the contractor designs the project, constructs it, and operates it for a period of time, for example 20 years

Design phases: three to five sometimes distinct and sometimes overlapping efforts to produce design documents with each phase becoming more accurate and more complete. Includes programming, conceptual, schematic, design development, and construction documents

Detailed cost estimate: extensive estimate based on definitive design documents. Includes separate labor, material, equipment, and subcontractor quantities. Unit prices are applied to material quantity take-offs for every item of work

Developer: individual or company that improves property, land and buildings, in order to take a profit; see also *real estate developer*

Development: construction project which improves the value of bare land or renovates buildings; see also *real estate development*

Direct construction costs: labor, material, equipment, and subcontractor costs for the contractor, exclusive of any markups

Discount: amount of money in the future, or future value, that is reduced to reflect the present value

Dispute: a contract claim between the owner and the general contractor that has not been resolved

Distributable costs: see *indirect construction costs, general conditions (2)*, and/or *overhead*

Earned value: a technique for determining the estimated or budgeted value of the work completed to date and comparing it with the actual cost of the work completed. Used to determine the cost and schedule status of an activity or the entire project, also known as earned value method or earned value approach

Earned value indices: formulas and ratios to compare estimated cost and schedule to actual cost and schedule utilizing the third earned value curve

Economy of scale: construction items cost less per each when their quantity is increased, such as 1,000 hotel rooms versus ten hotel rooms, and the opposite when quantities are decreased

Eighty-twenty rule: on most projects, about 80% of the costs or schedule durations are included in 20% of the work items; also known as Pareto's 80-20 rule

Equipment ledger: list of construction equipment including purchase price, amount of depreciation, book values, maintenance history, rental rates, and project assignments

Estimate schedule: management document used to plan and forecast the activities and durations associated with preparing the cost estimate. Not a construction schedule

Experience modification rating: a factor that reflects the construction company's past safety accident claims history. This factor is used to increase or decrease the amount charged for the company's workers' compensation insurance premium rates

Federal income tax: percent charged by the IRS to individuals and companies based on profit, which is calculated after subcontracting expense from revenue or income

Fee: contractor's income after direct project costs and jobsite general conditions costs are subtracted from revenue. Generally includes and is also known as *home office overhead* and *profit*

Field engineer: similar to the project engineer except with less experience and responsibilities. May assist the superintendent with technical office functions

Field question: see *request for information*

Final completion: the stage of construction when all work required by the contract has been completed

Final lien release: a lien release issued by the general contractor to the client or by a subcontractor to the general contractor at the completion of a project indicating that all payments have been made and that no liens will be placed on the completed project

Financial manager: see *chief financial officer*

Financial ratios: comparison of various entries in the balance sheet and income statement and other financial reports in order to provide the business owner and other stakeholders, such as the bank and bonding company and stockholders, with a more detailed picture of the financial strength of a company

Fixed overhead: portion of home office overhead that is spent or committed to be spent, regardless of company volume such as the chief executive and chief financial officers' wages and office building rent

Forecast: monthly cost estimate and report prepared by the project manager for upper-management which combines costs incurred with costs to go and results in a new estimated fee. Each monthly forecast should be more accurate than the last

Foreman: direct supervisor of craft labor on a project

Foreman work packages: see *work packages*

Front-loading: a tactic used by a contractor to place an artificially high value on early activities in the schedule of values to improve cash flow

Future value: forecasted worth an object has in the future depending on anticipated inflation rate; see also *present value*

General conditions (1):   a part of the construction contract that contains a set of operating procedures that the owner typically uses on all projects. They describe the relationship between the owner and the contractor, the authority of the client's representatives or agents, and the terms of the contract. The AIA general conditions document A201 is used by many clients and architects

General conditions (2):   indirect costs, whether in the home office or at the jobsite, which cannot be attributed solely to any direct work activities

General contractor: the party to a construction contract who agrees to construct the project in accordance with the contract documents; employs direct craft labor and subcontractors

General ledger: series of accounting documents and spreadsheets including the income statement and balance sheet

General liability insurance: protects the contractor against claims from a third party for bodily injury or property damage

General partner: individual or group of individuals amongst a group of 'partners' which are ultimate decision-makers; every partnership must have at least one general partner; see also *partnership* and *limited partner*

Guaranteed maximum price contract: a type of open-book cost-plus contract in which the contractor agrees to construct the project at or below a specified cost and potentially share in any cost savings

Home office: main office location, usually where the CEO and CFO work

Home office overhead: combination of fixed and variable costs associated with operation of the home office and the company, usually not assignable to individual construction projects; see also *general conditions (2)*

Income statements: financial report prepared by the CFO and potentially with the assistance of a CPA which begins with total revenue received, subtracts all expenses incurred at both the home office and the jobsite, and results in a gross profit bottom line from which taxes are deducted to produce the net profit

Income tax: mandatory financial contributions individuals and companies make to fund the government at various levels: Federal, state, county, and city

Indirect construction costs: expenses indirectly incurred and not directly related to a specific project or construction activity, such as *home office overhead*; see also *general conditions (2)*

Inflation: concept that prices increase in time, things will cost more in the future, usually reflected as a percentage; related to *time value of money*

Integrated project delivery method: fairly new contracting method where the project owner, architect, and general contractor all sign the same contract agreement and share risks equally in financial, safety, schedule, and quality performance

Interest: amount of money one party pays to another in order to use their dollars

Internal Revenue Service: Federal Government agency responsible for implementing the tax laws established by congress

Investment ratios: see *financial ratios*

Invitation to bid: a portion of the bidding documents soliciting bids for a project; also known as instructions to bidders or *request for quotation*

Job cost history report: monthly report of periodic and cumulative costs charged to a construction project organized by cost code

Jobsite general conditions costs: field indirect costs which cannot be tied to an item of work, but which are project-specific, and in the case of cost reimbursable contracts are considered part of the cost of the work; see also *general conditions (2)*

Jobsite layout plan: two-dimensional plan of the jobsite often authored by the project superintendent that includes locations for hoisting and trailers and site safety and storm water control, among others; see also *site logistics plan*

Joint venture: a contractual collaboration of two or more parties to undertake a project

Journal entry: accounting transaction of moving estimated and actual costs from one cost code to another

Just-in-time delivery of materials: a material management approach in which supplies are delivered to the jobsite just in time to support construction activities; this minimizes the amount of space needed for on-site storage of materials

Labor benefits: voluntary costs paid by contractors on top of employee wages including union dues, vacation, medical insurance, training, retirement, and others

Labor burden: addition of *labor benefits* plus *labor taxes*

Labor taxes: mandatory costs paid by contractors on top of employee wages including FICA, workers' compensation insurance, state and Federal unemployment insurance, and Medicare

Last planner: individual or group of individuals ultimately responsible for getting the work accomplished, both on the design side and construction side, which is usually a craft foreman or subcontractor

Lean construction: process to improve costs and eliminate waste incorporating efficient methods during both design and construction; includes value engineering and pull planning

LEED: Leadership in Energy and Environmental Design; a measure of sustainability administered by the United States Green Building Council, usually associated with receipt of a LEED certificate

Letter of intent: a letter, in lieu of a contract, notifying the contractor that the client intends to enter into a contract pending resolution of some restraining factors, such as permits or financing or design completion; sometimes allows limited construction or procurement activities to occur

Leverage: economic tool of applying the minimum amount of an investor's own money to get the most benefit, often by borrowing from others

Lien: a legal encumbrance against real or financial property for work, material, or services rendered which added value to that property

Lien release: a document signed by a subcontractor or the general contractor releasing its rights to place a lien on the project for a stipulated financial sum

Life-cycle cost: the sum of all acquisition, operation, maintenance, use, and disposal costs for a product over its useful life

Limited partner: individual or group of individuals who are equity owners but not responsible for management activities. Their financial risk in the company is limited to the amount of their initial investment

Liquidated damages: an amount specified in the contract that is owed by the contractor to the client as compensation for financial damage incurred as a result of the contractor's failure to complete the project by the date specified in the contract

Long-form purchase order: a contract for the acquisition of materials that is used by the project manager or the construction firm's purchasing department to procure major materials for a project; see also *purchase orders*

Long-term asset: item of value such as real estate and/or major construction equipment which is held longer than one year

Long-term financing: see *permanent loan*

Long-term liability: debt which will take longer than one year to pay off such as a loan on an office building or major piece of construction equipment

Lump-sum contract: a contract that provides a specific price for a defined scope of work; also known as a fixed-price or stipulated-sum contract

Markup: percentage added to the direct cost of the work to cover such items as overhead, fee, taxes, contingency, and insurance

MasterFormat: a 16-division numerical system of organization developed by the Construction Specifications Institute that was used to organize contract specifications and cost estimates. Today a new 49 CSI division is more commonly used

Materialman's notice: a notice sent to the client that a supplier will be delivering materials to the project and reserves their right to lien the property if payment is not made

Material supplier: vendor who provides materials but no on-site craft labor

Modified accelerated cost recovery system: depreciation system recognized by the IRS to allow increased equipment and real estate deductions during the early life of an asset

Monthly project manager's cost and/or fee forecast: jobsite financial report generated by the project manager which considers costs to date and adds projected costs to go for each estimate line item, coming to a bottom-line total which will translate into a revised fee forecast

Net cash flow: either a positive or negative flow of cash resulting from the deduction of expenses paid, such as wages to craftsmen and subcontractor payments, from payments received from the client, incorporating effects of retention. Contractors endeavor to have a positive net cash flow

Non-reimbursable costs: costs expended by a contractor which must be absorbed in the fee and not considered job cost; associated with open-book negotiated projects

Notice to proceed: written communication issued by the owner to the contractor, authorizing the contractor to proceed with the project and establishing the date for project commencement

Occupational Safety and Health Administration: Federal agency responsible for establishing jobsite safety standards and enforcing them through inspection of construction work sites

Officer-in-charge: general contractor's principal individual who supervises the project manager and potentially the superintendent and is responsible for overall contract compliance

Off-site construction: prefabrication of building modules or systems improving on-site cost and schedule performance; also known as off-site prefabrication

Operation and maintenance manuals: a collection of descriptive data needed by the client to operate and maintain equipment and materials installed on a project

Overbilling: requesting payment for work that has not been completed

Overhead: expenses incurred that do not directly relate to a specific project, for example, rent on the contractor's home office; see also *general conditions (2)*

Overhead burden: a percentage markup that is applied to the total estimated direct cost of a project to cover overhead or indirect costs

Partnership: legal form of business ownership where a group of individuals contribute equity and may or may not share management duties; see also *general partner* and *limited partner*

Payback period: see *cost recovery period*

Payment bond: guarantee by a surety that a contractor will pay its subcontractors and suppliers, also known as labor and material payment bond

Payment request: document or package of documents requesting progress payments for work performed during the period covered by the request, usually monthly, also known as a pay estimate

Percentage of completion accounting method: accounting method associated with larger contractors and longer-term projects where expenses and revenue are recognized when work is accomplished, often ahead of when payments are made and received

Performance bond: a guarantee by a surety that a contractor will complete the work it has been contracted to do

Permanent loan: long-term or take-out loan given to a developer by banks and other investment institutions after construction is complete

Plugs: general contractor's early cost estimates or allowances for subcontracted scopes of work

Preconstruction agreement: a short contract that describes the contractor's responsibilities and compensation for preconstruction services

Preconstruction services: services or activities that a construction contractor performs for a project client during design development and before construction starts; includes estimating, scheduling, and constructability reviews

Pre-proposal conference: meeting of potential contractors with the project client and architect. The purpose of the meeting is to explain the project, the negotiating process and selection criteria, and solicit questions regarding the design or contract requirements; similar to a pre-bid conference

Prequalification of contractors: investigating and evaluating prospective contractors based on selected criteria prior to inviting them to submit bids or proposals

Present value: today's worth of an investment made in prior years or anticipated to be made in future years and adjusted for inflation or interest

Profit: the contractor's net income after all expenses have been subtracted

Procurement methods: a project owner will choose to employ a general contractor either by soliciting competitively priced bids or negotiated proposals, and general contractors will do likewise with subcontractors and suppliers

Pro forma: spreadsheet and formula for real estate developers to compare the projected cost of a construction project to its value utilizing an expected revenue stream

Progress payments: periodic (usually monthly) payments made during the course of a construction project to cover the value of work satisfactorily completed during the previous period

Project control: methods the project team utilizes to anticipate, monitor, and adjust to risks and trends in controlling cost, schedule, quality, and safety

Project engineer: project management team member who assists the project manager on larger projects. The project engineer is usually more experienced and has more responsibilities than the field engineer, but less than the project manager. The project engineer is responsible for management of technical issues on the jobsite; see also *cost engineer*

Project labor curve: a plot of estimated labor hours or crew size required per month for the duration of the project

Project management: application of knowledge, skills, tools, and techniques to the many activities necessary to complete a project successfully

Project management close-out: end of project completion of paperwork and documentation including as-built drawings, O&M manuals, test reports, prevailing wage rate schedule, and many others as prescribed in the prime agreement

Project management organization: the contractor's project management group headed by the officer-in-charge, including field supervision and staff

Project manager: the leader of the contractor's project team who is responsible for ensuring that all contract requirements are achieved safely and within the desired budget and time frame; usually supervises field office staff, including project engineers, cost engineers, and/or jobsite cost accountants

Project manual: a specification volume that also may contain contract documents such as instructions to bidders, bid form, general conditions, special conditions, and/or the geotechnical report

Project team: individuals from one or several organizations who work together as a cohesive team to construct a project

Property tax: fee or tariff charged by a state or local government on a private owner levied as a percentage of the value of real estate; used to pay for roads and schools and other municipal services

Public-private-partnership: a construction delivery method in which a public agency partners with a contractor or developer to reduce costs and lawsuits and ultimately save taxpayer money

Pull planning: scheduling method often utilizing stickie-notes where milestones of each design or construction discipline are established and the project is scheduled backwards with the aid of short-term detailed schedules; a tool of *lean construction*

Punch list: a list of items that need to be corrected or completed before the project can be considered completed

Purchase orders: written contracts for the purchase of materials and equipment from suppliers; see also *long-form purchase order* and *short-form purchase order*

Quality control: process to assure materials and installation meets or exceeds the requirements of the contract documents

Quantity take-off: one of the first steps in the estimating process to measure and count items of work to which unit prices will later be applied to determine a project cost estimate

Ratios: also known as investment ratios; see *financial ratios*

Real estate developer: a person or entity that engages in real estate development usually with the intention of making a profit

Real estate development: process to improve land or buildings to change use and/or increase value

Reimbursable costs: contractor costs incurred on a project that are reimbursed by the client on a negotiated project. The categories of costs that are reimbursable are specifically stated in the contract agreement

Request for information: document used to clarify discrepancies between differing contract documents and between assumed and actual field conditions; also known as *field question*

Request for proposals: document containing instructions to prospective contractors regarding documentation required and the process to be used in selecting the contractor for a project

Request for qualifications: a request for prospective contractors or subcontractors to submit a specific set of documents or responses to demonstrate the firm's qualifications for a specific project

Request for quotation: a request for prospective contractors to submit a quotation for a defined scope of work; also known as instruction to bidders and/or information to bidders

Retention: a portion withheld from progress payments for contractors and subcontractors to create an account for finishing the work of any parties not able to or unwilling to do so; also known as retainage

Revenue: total money received by contractors before cost deductions are made

Risk analysis: identification and acknowledgement of potential downsides of a decision or endeavor, such as an investment or acceptance of a construction contract, and the development of methods to mitigate those risks

Rough-order-of-magnitude cost estimate: a conceptual cost estimate usually based on the size of the project. It is prepared early in the estimating process to establish a preliminary budget and decide whether or not to pursue the project

Sales tax: percentage fee charged by a state or local government for sale of materials, services, or real estate used to pay for schools, roads, libraries, and other similar municipal facilities or services

S corporation: legal form of business organization usually associated with smaller or start-up companies, also known as S corp

Schedule of values: an allocation of the entire project cost to each of the various work packages or systems required to complete a project. Used to develop a cash flow curve for an owner and to support requests for progress payments; also serves as the basis for AIA document G703, which is used to justify pay requests

Self-performed work: field construction work performed by the general contractor's work force rather than by a subcontractor

Short-form purchase order: purchase order document used on project sites by superintendents to order materials from local suppliers; see also *purchase order*

Short interval schedule: schedule that lists the activities to be completed during a short interval (two to four weeks). Also known as a look-ahead schedule; used by the superintendent and foremen to manage the work

Site logistics plan: pre-project planning tool created often by the general contractor's superintendent which incorporates several elements including temporary storm water control, hoisting locations, parking, trailer locations, fences, traffic plans, etc.; see also *jobsite layout plan*

Sole proprietorship: legal form of business ownership where an individual's personal assets and income are mixed with the company's. Usually associated with smaller construction companies and consultants

Special conditions: a part of the construction contract that supplements and may also modify, add to, or delete portions of the general conditions; also known as supplementary conditions

Specialty contractors: construction firms that specialize in specific areas of construction work, such as painting, roofing, or mechanical; see also *subcontractors*

Specifications: see *technical specifications*

State income tax: percentage fee charged by most states to individuals and legal entities based on revenue received less expenses paid which equals their profits

Subcontractors: specialty contractors who contract with and are under the supervision of the general contractor

Subcontracts: written contracts between the general contractor and specialty contractors who provide jobsite craft labor and usually material for specialized areas of work

Submittals: shop drawings, product data sheets, and samples submitted by suppliers and subcontractors for verification by the design team that the materials intended to be purchased for installation comply with the design intent

Substantial completion: state of a project when it has been sufficiently completed such that the owner can use it for its intended purpose

Summary schedule: abbreviated version of a detailed construction schedule that may include 20 to 30 major schedule activities or milestones

Sunk costs: expenditures made in the past that cannot be recovered

Superintendent: individual from the contractor's project team who is the leader on the jobsite and who is responsible for supervision of daily field operations on the project including quality and safety

Superintendent diary: see *daily job diary*

Suppliers: companies which provide materials to the construction project for installation by others, either the general contractor or subcontractors. The supplier may or may not engage in off-site material fabrication

Supply chain material management: recognition that costs start at the material fabrication facility and true savings can be saved earlier in the process at that location rather than later during construction installation

Sustainability: broad term incorporating many green-building design and construction goals and processes including *LEED*

Target value design: top-down design process starting with the client's total project budget, dividing it up amongst cost elements, and then preparing the design to fit each individual budget line item

Technical specifications: a part of the construction contract that provides the qualitative requirements for a project in terms of materials, equipment, and workmanship

Third-tier subcontractor: a subcontractor who is hired by another firm that has a subcontract direct with the general contractor

Time and materials contract: a cost-plus contract in which the client and the contractor agree to a labor rate that includes the contractor's profit and overhead. Reimbursement to the contractor is made based on the actual costs for materials and the agreed labor rate times the number of hours worked

Time value of money: economic study comparing past, present, and future values of money using a series of formulas and spreadsheets and relying on several factors such as interest rates, inflation rates, investment years, and other variables

Total quality management: philosophy of quality control which starts at the top of an organization, at the CEO level, and transgresses all the way through to jobsite management, supervision, and craftsmen

Traditional project delivery method: a delivery method in which the client has a contract with an architect to prepare a design for a project; when the design is completed, the client hires a contractor to construct the project

Unconditional lien releases: see *final lien release*

Unresolved change orders: see *claim*

Value engineering: a study of the relative values of various materials and construction techniques to identify the least costly alternatives without sacrificing quality or performance, and/ or improving life-cycle costs

Variable overhead: home office costs that change with respect to company volume, such as insurance costs; can also be applied to some jobsite indirect costs

Warranty: guarantee that materials furnished are new and able to perform as specified and that all installed work is free from defects in material or workmanship

Warranty management: process to handle and log warranty claims and repairs

Weekly labor report: cost report document generated by a contractor's home office on a weekly or twice monthly basis that records periodic and cumulative jobsite direct and indirect labor costs against the original estimate and assigned cost codes

Work breakdown structure: a list of significant work items that will have associated cost or schedule implications

Work package: a defined segment of the work or system required to complete a project

Workers' compensation insurance: insurance that protects the contractor from a claim due to injury or death of an employee on the project site. Rates are reflective of a contractor's safety history; see also *experience modification rating*

Index

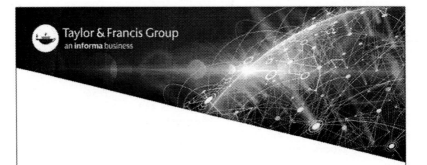